Lake George New York

CANADA'S
FRONT DOOR

—

AMERICA'S
BACK DOOR

Edward J. Dodge

HERITAGE BOOKS
2012

HERITAGE BOOKS
AN IMPRINT OF HERITAGE BOOKS, INC.

Books, CDs, and more—Worldwide

For our listing of thousands of titles see our website
at
www.HeritageBooks.com

Published 2012 by
HERITAGE BOOKS, INC.
Publishing Division
100 Railroad Ave. #104
Westminster, Maryland 21157

Copyright © 2012 Edward J. Dodge

Other Heritage Books by the author:
Relief is Greatly Wanted: The Battle of Fort William Henry

All rights reserved. No part of this book may be reproduced or transmitted in any form or by any means, electronic or mechanical, including photocopying, recording or by any information storage and retrieval system without written permission from the author, except for the inclusion of brief quotations in a review.

International Standard Book Numbers
Paperbound: 978-0-7884-5393-9
Clothbound: 978-0-7884-9314-0

DEDICATION

To the fallen. May they rest in peace with the warm sun to their face and a fair wind to their back.

Table of Contents

Dedication iii

List of Illustrations vii

Acknowledgements ix

Foreword xi

Chapters:

1. Lake George—The Door 1
   This article appeared in the Spring 2007 *Fort George Advice*

2. The Battle of Lake George, 8 September 1755 3
   This article appeared in the Spring 2005 *Fort George Advice*

3. Samuel Blogett's Map of the Battle of Lake George 7
   This article appeared in the Fall 2005 *Fort George Advice*

4. Blogett's American Published Map of the Battle 9
   This article appeared in the Spring 2006 *Fort George Advice*

5. Blogett's British Published Map of the Battle 13
   This article appeared in the Fall 2006 *Fort George Advice*

6. King Hendrick 17
   This article appeared in the Spring 2004 *Fort George Advice*

7. Ephraim Williams 23
   This article appeared in the Fall 2004 *Fort George Advice*

8. Major General Seth Pomeroy 27
   This article appeared in the Fall 2007 *Fort George Advice*

9. A Second Look 33
   This article appeared in the Spring 2008 *Fort George Advice*

10. Timothy Ruggles 37
    This article appeared in the Fall 2008 *Fort George Advice*

11. Once Again a Return to the Battlefield 41
    This article appeared in the Spring 2007 *Fort George Advice*

12. Jared Sparks 47
    This article appeared in the Winter/Spring 2009 *Fort George Advice*

13. The 253 Year Old Mystery 53
    This is the first appearance of this article

14. The French and the Lake 57
    This article appeared in the Spring/Summer 2010 *Fort George Advice*

About the Author 65
Bibliography 66

List of Illustrations

Lake George in a late summer storm—looking North from the South shore xii

Sir William Johnson by Francis Whiting Halsey 2

Blodgett's Map — Boston 1755 10

Samuel Blodgett's Map — London 1756 14

King Hendrick, Mohawk Chief 18

Jared Sparks by Rembrandt Peale 48

Jared Sparks' sketch of Fort George, August 1830 50

Head of Lake George 51

ACKNOWLEDGEMENTS

This book would never have gotten written without the help and encouragement of friends and associates.

Carole Tracy who pushed me to do it; Mary Lu Metz who read the draft and stated "it's a book"; Nadine Battaglia ever the staunch supporter; Tim Todish, author, historian and reenactor; Brent C. Kemmer, educator, historian, author and reenactor; Ian McCulloch, historian and author; Chris Fox, Curator – Pell Research Center, Fort Ticonderoga; Dean R. Snow, Chair – Department of Anthropology, Penn State; Nancy Fay, Library and Archives Canada; Linda Hall, Archives Assistant and Sylvia Brown, College Archivist, both of Williams College; Mary Haegert and Heather Cole, Houghton Library, Harvard University; Christopher Wm. Linnane, Harvard Art Museum; Nancy Maliwesky, Director – American Pomeroy Historical Genealogical Association and co-author of the chapter on General Pomeroy; Herman C. Brown, President – Lake George Battlefield Park (Fort George) Alliance, whose Newsletter [Fort George Advice] previously published all these chapters as individual stories with the exception of "The 253 Year Old Mystery"; Leslie Wolfinger and Debbie Riley, Heritage Books; and The Photo Resource Center, Springfield, Illinois.

FOREWORD

In a letter dated 25 February 1897, General Lewis M. Peck, Retired, U. S. Volunteers, wrote to an acquaintance that "I apprehended that few of his kin, wisely or comprehensively, can note how all (of) that which occurred at Lake George, insensibly has moulded the present civilization of this land".

The General was referring to one of the defenders of Lake George, possibly all of the defenders.

In a letter written by one of those defenders from his tent on 11 September 1755, Colonel Seth Pomeroy wrote to his sister-in-law about the battle and the death of her husband (his brother) and the aftermath. He told her of the troops bringing in a wounded man who had been sitting where they found him since the 8th of September, the day of the battle, with the back of his skull cut off and his brain exposed,

There was no way that man or those who found him or the man that wrote about him and the death of his brother could ever know that battle was shaping the civilization of this land.

No soothsayers or visions in the clouds, only the smell of blood, the smell of death, cries of the wounded and sounds of mourning occupied their minds at that time. History was being made, but they did not know or care. Only for their own time did they care.

Lake George in a late summer storm—looking North from the South shore.
Photo by author.

LAKE GEORGE - THE DOOR

What do you know about Lake George? Today it is a major resort area and in the past it has seen three major military battles, if not four.

The lake is in an old glacier trough. The mountains on either side are primarily composed of granite that dates from the Paleozoic age which was some three hundred thousand years ago. The lake is considered to be in the Adirondack Mountains. To the west the mountains are some 2200 feet high and to the east some 2600 feet high. As to whether they were formed by volcanic action or earth quake as they are just south of the St. Lawrence Fault Line is probably up for debate.

The lake is known as the tail of Lake Champlain. It is orientated south to north, flowing through the narrows or portage, dropping some 250 feet into Lake Champlain. The glacier that created the lake receded some 10,000 years ago. The lake is 32 miles long, 1 to 3 miles wide with 109 miles of shore line. It contains some 44 square miles of water, 300 islands and at its deepest is 195 feet deep. It's color is deceptive. Up close it has a brownish hue caused by the heavy iron deposits in the area. At a distance it appears blue, green or a combination of the two colors.

Artifacts recovered from archaeological excavations at the south end of the lake which is square shaped, reflect that man camped on the shores of the lake as early as 5000 years ago. It appears that these were fishing camps and of short duration. Summer is nice at the lake although it can get some violent storms. Winter on the other hand is extremely severe with temperatures well below freezing if not below zero degrees Fahrenheit.

The area where the lake is located was known to the Native American tribes of the northeast and Canada as The Great Wilderness. The lake has had many names from various sources. To the Native Americans it was Andiatarocte; to the French it was Lac St. Sacrament; to the early settlers it was Iroquois Lake and to the British Army it was one of the Little Lakes (Loon Lake, Schroon Lake, etc.). Sir William Johnson named it Lake George in honor of his king, King George. The writer James Fenimore Cooper referred to it as Lake Horican in his novel "The Last of The Mohigans".

Time moved forward as did the history of the lake. On 8 September 1755, The south end of the Lake was the site of a major battle between the French and British, again in March of 1757 and again in August 1757. The fourth major military engagement was British Major Carleton's Raid in October 1780 during the Revolutionary War.

The town of Lake George was originally the town of Caldwell. The village of Lake George became an incorporated village in 1903. Further down the lake, in August 1910 six months after the Boy Scouts of America was founded, would be the very first American Boy Scout Summer Camp. It was held at Silver Bay YMCA Camp. One hundred and twenty-five boys attended.

Now these are just a few of the things to know about Lake George. For further information, the reader is referred to your most favorite Encyclopedia; "New France and New England" by John Fiske, Heritage Books; "Notes on the History of Fort George" by B. F. DeCosta; "Fort William Henry, A History, Lake George, New York" by Stanley M. Gifford; "Relief is Greatly Wanted" by E. J. Dodge, Heritage Books - just to name a few - or stop by and visit with Mr. Todd DeGarmo or browse the collections at the Folk Life Center, Crandall Public Library, Glens Falls, New York. ■

This article appeared in the Spring 2007 *Fort George Advice*

Sir William Johnson by Francis Whiting Halsey.
Public domain.

THE BATTLE OF LAKE GEORGE
8 SEPTEMBER 1755

The Battle of Lake George to the French and Indian War enthusiast and military historian is worth discussion, possibly debate. At the time it was the largest battle fought in New England, possibly North America.

Numerous articles have been written about the battle with a multitude of fiction and error that one wonders where all the misinformation came from. The most accurate telling of events on that date is found in "A Prospectus - Plan of the Battle near Lake George on the Eighth Day of September 1755. With an Explanation thereof, containing a full, though short, History of That Important Affair, by Samuel Blodget occasionally at the Camp, when the Battle was fought."

Samuel Blodget was born 1 April 1724. During the battle he was a sutler to a New Hampshire regiment. When not doing that he was a farmer, manufacturer and canal builder. He built the canal around Amoskeag Falls on the Merrimac River in New Hampshire. He served as a sutler during the Revolutionary War. He died in September 1807. His Battle near Lake George perspective was published in Boston by E. B. O'Callahan in late 1755 and a year later in London.

To bring some form of focus to the numerous articles on the subject, one major writer has indicated that Baron Dieskau was killed during the battle, which is complete fiction. Writers and historians also miss the fact that this was just one battle in a world war as England was fighting in India, the West Indies and North America at the same time. The battle had its origin in Europe.

Baron Dieskau, the French Commander during the battle, was born Baron Ludwig August Dieskau in Saxony in 1701. There is some argument as to whether he was Polish or German. This is a moot point as Saxony was claimed by both countries at one time or another.

He chose to join the French army and was apparently known as Jean Armand Baron de Dieskau to the French. He was a protégé of Marshal de Saxe, serving as an Aide de Camp until 1744 when he was promoted to Colonel of Cavalry. By 1747 he was a Major General and military Governor of Brest, France. One writer has stated that Dieskau was an aide to Napoleon, who was born two years after Dieskau died.

On 1 March 1755 he was appointed Commander of French regular battalions being sent to Canada, arriving at Quebec in June 1755. By July 1755 he was at St Fredric (Crown Point). It was there that intelligence was received that the British were planning to attack St Fredric.

Two things were decided. One was to build Fort Carillon, now know as Fort Ticonderoga, and the second was to move to intercept the British force. The plan developed to move on the British was based on phony information from a British prisoner that Fort Lydus (Fort Lyman), at the Great Carrying Place on the Hudson River, was under construction and under strength. Dieskau headed south with a force of 3500 composed of Regulars, La Marine, Militia and Indians, commonly referred to as "Hurons."

By early September Dieskau was near Fort Lyman and discovered that, although it was still under construction, the fort was much larger than he had been told and well defended by cannon. His Indian allies vehemently turned down an attack on the fort in view of the cannon, which they did not like to fight against. In the same time frame his scouts captured two teamsters from whom he learned of William Johnson's camp at Lake George. Dieskau finally allowed his Indians to talk him into attacking Johnson's camp. His Indians had come for easy scalps, plunder and "the brew".

Dieskau proceeded northward along the military road from Fort Lyman to Lake George and set up camp roughly two and a half to three miles south of Lake George.

In the early a.m. of 8 September, Dieskau's scouts informed him of a British force moving south on the military road and an ambush was set which was located some two and a half miles from the lake. According to Blodget the ambush was set with an estimated 2000 men. On one side of the road they were protected and covered by a thick growth of brush and trees and on the other by a slight rise covered with rocks, trees and shrubs about chest high.

The British force under William Johnson was in reality provincial, although Johnson himself could be considered British. He was born in County Meath, Ireland in 1715. He arrived in North America in 1738 to manage a land tract that his uncle, Admiral, Sir William Peter Warren had purchased in the Mohawk Valley. Johnson gained quick acceptance with the Mohawks as he treated them fairly and honestly and became quite fluent in their language. They adopted him, naming him "WANAGHI-YAGEY" or "The Great Brother," and eventually made him a Sachem (leader of the tribe or nation). He married a German girl who gave him two sons. She died at a young age. He later married Mary (Molly) Brant, sister of the War Chief Joseph Brant in an Indian ceremony. Johnson never acknowledged her as his lawful wife even though they had eight children together and she stayed with him until his death.

In 1755 Governor Shirley of Massachusetts appointed Johnson commander of a force to attack St Fredric. Shirley was the generally acknowledged leader of the New England colonies or plantations even though they separately had their own governor. The force was to be composed of provincials, a British officer or two and Mohawks. It was believed the Mohawks would be of invaluable service in providing scouts and hunters. The plan had the concurrence of the various governors with only one proviso, that General Phineas Lyman of Connecticut would be Johnson's Number Two.

Johnson was given the honorary title of General even though he had never been taught the art of warfare or fought in any battles.

His command consisted of eight provincial regiments and some 250 to 300 Mohawks. It formed at Albany, New York.

He moved his command to Fort Lyman arriving on July 27th. By the 28th of August he was at Lake George preparing to move on St Fredric. His plans were thrown in disarray when his scouts informed him of the French construction at the carry between Lake Champlain and Lake George and that a large French force was near or at Fort Lyman. To this end he sent a messenger to Fort Lyman who it was later learned was killed by French Indians.

Johnson directed Colonel Ephraim Williams (see "Ephraim Williams" in the Fall 2004 newsletter) and his regiment to lay out the foundations of what would become Fort William Henry on a bluff which was separated by a particularly nasty marsh from the west end of his camp. He further directed Colonel Williams to lead a relief column of some 1200 men to the relief of Fort Lyman.

Blodget in his narrative makes a point of stating that the battle was a three-phase battle and not three separate battles.

Apparently after some confusion and rankling the relief column moved out from the camp at Lake George some two to three hours later than planned. As it proceeded down the military road the Mohawk scouts spotted the ambush and shots were fired. These shots alerted the column and a somewhat nasty fight occurred which saw the death of Chief Hendrick (see "King Hendrick" in the Spring 2004 newsletter), Colonel Williams and others.

The French had good position in that their cover was extremely good on one side and on the other, although not as good, they were on a low ridge which the provincials attempted to take in what became a hand-to-hand fight.

The provincials were outnumbered roughly two to one, yet did not fall back in panic; rather they put up a spirited defense and fell back in a well-organized retrograde maneuver orchestrated by Colonel Nathaniel Whiting, Colonel Williams' Number Two. The effect was more of the French were killed than the provincials according to a statement made by Dieskau. The ambush took place some two and a half miles from Johnson's camp at about nine in the morning. According to sources from Fort Lyman, they heard a heavy volume of firing between nine and ten in the morning even though they were roughly fourteen miles from the site of the battle. This would indicate the weather was clear, bright, sunny, with a wind blowing south off the lake.

Blodget specifically notes that Hendrick was on horseback as he was on in years and corpulent (stocky, not fat) and dressed as an Englishman.

The column retreated down the road as it was the only clear space in which they could move with order. One source has indicated that Whiting had the front rank fire and fall back, then the second rank would fire and fall back with each succeeding rank doing the same. Another source indicates that as the column came within view of the camp, some three hundred men from the camp joined them to provide additional firepower. Blodget made no mention of such an action and no other source supporting this statement could be found.

When the French were within 140 yards of the camp they brought their fire to bear on the camp. The French regulars whom Blodget estimated at somewhere between four and five hundred were in platoon on line and firing in that order.

Johnson's camp had a defense line of hastily cut trees which were cut down and added to already downed trees once the firing of the ambush was heard. It was not a continuous line and as all the undergrowth had been cleared away the only cover was the trees.

The French on their side had trees, shrubs and rocks, as well as a large wind fallen tree some 300 feet in front of the camp and a small ridge some 250 feet in front and uphill of the camp from which they could harass and gall the camp.

Johnson had three heavy cannon and a field piece near the center of his defensive line. These did do some damage to the French regulars before they took cover in the forest; however, the heavy cannon fired only 12 to 15 rounds and the field piece no more than two rounds. Two mortars on the right of Johnson's line fired two rounds into the marsh that separated the camp and the bluff as it was thought there were Indians there. Other cannon though out the camp were never used.

The provincials laid down a deadly fire against the French. The provincials were men who could bark a squirrel at a hundred yards. Their weapons protected them, fed their families and won shooting matches.

The French held the high ground with the advantage of good cover and concealment. Yet as they were firing downhill their shots had a tendency to go high. The provincials were shooting up hill and their shots had a tendency to drop and hit the intended target with devastating impunity.

Colonel Seth Pomeroy who was apparently a staff officer to Johnson, stated, "The Canadians and Indians helter-skelter, the woods full of them, came rushing with undaunted courage right down the hill upon us, expecting us to flee."

Dr. Thomas Williams (younger brother to Colonel Ephraim Williams) would later write, "It was the most awful my eyes ever beheld. There seemed to be nothing but thunder and lightning and perpetual pillars of smoke."

An unidentified soldier wrote, "The hailstones from Heaven were never much thicker than their bullets came."

The battle raged for some four to five hours until about four or five in the afternoon with the French finally pulling back. One writer has indicated that General Lyman who had taken over after Johnson was wounded in the leg and retired to his tent, had ordered a bayonet charge which cleared the French from the field. Blodget made no mention of this and I found no other source confirming this statement.

The battle had occurred on a line as the provincials faced the French, which was anchored at the lake's edge on the left and anchored at the marsh on the right. It was defended by some 1800 provincials and Mohawks with another 500 men spread around the camp as sentries. The total provincial force numbered about 2250. The French force numbered somewhere between 2000 and 3000 during the engagement.

Colonel Joseph Blanchard at Fort Lyman upon hearing the sounds of battle for over an hour in the morning decided to send a relief column of between two and three hundred men to assist. The column under the joint command of Captains McGinnis and Folsom arrived at the site of the French camp between four and five in the afternoon.

There they found about 500 of the French force (mostly Indians) who had deserted the battle at the lake, apparently to scalp and mutilate the dead and wounded from the morning ambush. The column attacked, driving them from the camp, with the flight of the French and Indians being so rapid many of the scalps they had taken were dropped and later recovered by the provincials. They pursued the French until twilight, killing an estimated 100 of them outright. Legend states their bodies were tossed into a pond, giving rise to the story "The Battle of Bloody Pond." Blodget made no mention of such an event, but possibly alludes to it in that the bodies if fifty French and Indians were found in one spot.

Enough ammunition, provisions and other items were recovered from the French camp that the total filled four to five wagons.

Both forces had a great deal of military talent. Dieskau had Captain Jacques Le Gardeur de St. Pierre who was in charge of his Indians composed of Abenaki, Nippisings and Caughnawagas (Canadian Mohawks); Captain Charles Chevalier de Raymond, Captain Jean Baptiste Rene Legardeur de Repantigny and a de Vassan. All army or LaMarine.

Johnson had General Lyman, Colonel Seth Pomeroy, Colonel Ephraim Williams, Colonel Nathaniel Whiting, Colonel Harris, Colonel Cockroft, Colonel Ruggles, Colonel Titcomb and Colonel Gutridge.

The only identified British Army officer was Captain William Eyre of the 44th (Irish) Regiment of Foot. He was a trained engineer and handled the artillery for Johnson. He was also responsible for the construction of Fort Lyman (Fort Edward) and Fort William Henry.

The battle was costly in wounded, missing and dead enlisted personnel, senior officers, junior officers and Indians. The provincials lost Chief Hendrick, Colonel Williams, Colonel Titcomb, Captains McGinnis and Folsom, an unnamed major, a Lieutenant Barron and several other unnamed company grade officers. The French lost Captain St. Pierre and eight unnamed LaMarine officers. Dieskau would later relate that most of his officers were killed. Blodget estimated at least 700 French and Indians were killed. The provincials lost 126 plus 20 Mohawks killed; 60 provincial were missing and 94 provincials were wounded as well as 12 Mohawks. Given the known strength on each side, this is roughly a 10% casualty rate for each side.

French prisoners were taken, among them Dieskau. There is no recorded information that any provincials were taken prisoner. Dieskau had been deserted by his troops. He had been shot three times in the leg and allegedly a fourth time in the groin. He directed his troops from a sitting position against a tree. Upon his capture he was taken to Johnson's tent where he was attended. In something of a turnabout, Johnson had to personally protect him from the Mohawks who were seeking revenge for Hendrick's death. Dieskau noted in his papers, which were recovered, that he was less than happy with his Indian allies. He wrote, "They drive us crazy from morning to night. There is no end to their demands. They have already eaten five oxen and as many hogs without counting the kegs of brandy they have drunk." He later told his Indians "to not take time to scalp the wounded or dead as they could kill ten men in the time it took to scalp one."

He told Johnson of his admiration of the provincials who he stated, "in the morning they fought like little boys about noon like men and in the afternoon like devils."

Blodget stated that the battle would be recorded as the greatest battle and victory on the New England annals.

It was and is still regarded as a major victory for the British and provincials.

What happened to the Principals? Dieskau was repatriated in 1763 and died in France in 1767. Johnson became a Baron and received a cash reward from the Crown. General Lyman had the fort named after him and faded into history as did all the colonels. Fort Lyman was later renamed Fort Edward. Captain Eyre became a Major, defended Fort William Henry against a French attack in March 1757 and later died at sea when the ship he was sailing home on was lost with all hands off the coast of Ireland in 1760.

There is one tale of the battle that is both true and wrong. One source has indicated that Robert Rogers and his Rangers took part in the "Battle of Bloody Pond." Not really. The Robert Rogers at the battle was born in Scotland

and was a Private in a New York Ranger company commanded by Captain Isaac Corsa of Westchester County, New York. The other Robert Rogers was a captain in charge of a Ranger company in Colonel Joseph Blanchard's New Hampshire regiment at Fort Lyman, out with a detachment of his men up the Hudson River at the time of the battle at the order of William Johnson.

There also has been some discussion as to the actual location of Bloody Pond. In 25 August 1761, Surveyor Archibald Campbell undertook the task of locating "the Pond called Bloody." He headed south from Fort George and located the pond some 144 chains south of the fort. A chain is 66 feet in length. Multiplying that by 144 gives a figure of 9504 feet or roughly 1.8 miles.

As battles go, the Battle of Lake George was great in its day, but no better or worse than any other. ■

REFERENCES

- *Relief is Greatly Wanted, The Battle of Fort William Henry*, Ed Dodge, Heritage Books, Bowie, MD
- *New France and New England*, John Fiske, Heritage Books, Bowie, MD
- *LaMarine, The French Colonial Soldier in Canada, 1745-1761*, Andrew Gallup & Donald E. Shaffer, Heritage Books, Bowie, MD
- *Fort William Henry 1755-1757, a History, Lake George, NY*, Stanley M. Gifford
- *The Summer in Paradise in History*, Warwick Stevens Carpenter, The Delaware & Hudson Co., Albany, NY
- *The Illustrated Columbia Encyclopedia*, Vols. 6, 11 &12, Columbia University Press, New York & London
- *Colonial Wars of North America 1512-1763*, edited by Alan Gallay, Garland Publishing, New York & London
- *The American Wars, A Pictorial History from Quebec to Korea*, Roy Meredith, The World Publishing Co., Cleveland & New York
- *The History of Roger's Rangers, Vol. I - The Beginnings*, Burt Garfield Loescher, Heritage Books, Bowie, MD
- *Prospective-Plan of the Battle near Lake George*, Samuel Blodget, E. B. O'Cahallan, Boston & London
- *Land Survey Field Book*, Archibald Campbell, Vols. 39-41, Specifically Vol. XL, Michro A4019-77 MU1 Reel 14, New York State Archives, Albany, NY
- Muster Rolls New York Provincial Troops 1755-1764, page 9, New York Historical Society Publications Fund Vol. XXIV, Albany, NY

This article appeared in the Spring 2005 *Fort George Advice*

Samuel Blodgett's Map of the Battle of Lake George

Blodgett's Prospective Plan of the Battle of Lake George, September 8th, 1755, is probably one of the few first-hand accounts, if indeed not the only one, of a major battle in early North America.

What made it more unusual is that Blodgett, who was an amateur engineer, left detailed notes of what took place.

Colonel H. Avery Chenoweth, Sr., USMCR, Retired, in the Combat Art vs. Illustration section of his book [Art of War], the first paperback edition copy-write by Barnes & Noble, 2003, wrote as follows: "For our purposes combat art is defined as art in and from actual observed or experienced battle, as opposed to historical battle art or illustration created from imagination. The latter two are valid in their own right and the result of after the fact research and reconstruction, quite often from eyewitness accounts. Illustration is always imaginary, often fanciful, and the spark of authenticity of the actual observer is absent."

Blodgett's own print (combat art) was executed as an engraving by Thomas Johnson who gained notoriety as a Japanner. Whether the print was done on copper is not really known, but the likelihood is definitely there.

A year after it appeared in America in 1755 it was printed in London. The major differences between the two is that the American version shows a map on the Hudson River across the top of the page with Phase one and Phase two of the battle side by side from left to right and a dedication to His Excellency William Shirley, Esqr. in the print's lower left corner.

The American version shown in figure 1 comes from the holdings of the New York Public Library. The London version which is better known as shown in figure 2, comes from the Papers of Sir William Johnson, University of the State of New York, Division of Archives & History, Vol. IX.

The unusual aspect of the Prospective Plan of the Battle is that the Prospective proper is a detailed explanation of number coded points on the map. It is quite unusual to see a whole battle set forth in such fashion and definitely takes the reader right to that point of the action (see "The Battle of Lake George - 8 September 1755" in the Spring 2005 newsletter).

There are variations on both maps due to copies having been made by different engravers of that time and later. A map is discovery and can establish or skew history/

REFERENCES:

Antiques Magazine, Vol. LXXX #5, November 1961

Art of War, Col. H. Avery Chenoweth, Sr., USMCR, Ret., Barnes & Noble, 1st Paperback Edition, 2003

Papers of Sir William Johnson, University of the State of New York, Division of Archives & History, Vol. IX

Article appeared in Fall 2005 Fort George Advice

Blodgett's American Published Map of the Battle of Lake George

Modern man views a map as one of two things. First, as a way to find the shortest most direct route to where he wants to go. Secondly, as a small or large multi colored sheet of paper with a lot of symbols that he does not understand.

Historically, maps are about as old as man. Our ancestors using a finger or twig would draw lines in dirt, sand, snow, or scratch on rock with another rock to show someone the way from where he or she was to where he or she wanted to go. The oldest known map which still exists is a clay tablet from Mesopotamia showing a man's estate in great detail, dating from about 2800 BC.

A map by definition is a selective, symbolized and generalized picture on a much reduced scale of some spatial distribution of a large area, usually the earth's surface, as seen from above.

A Planimatric map does not show relief. A Topographic map shows relief (mountains, valleys, rivers, etc.). To accomplish this hydrographic symbols and symbols for man made objects are used and are basically modern in concept.

The earliest known colonial map from North America is a woodcut of Boston done by Mr. John Foster in 1677. Most maps of North America were done by English, French, Dutch or Spanish cartographers. It is believed that the English reached North America in the vicinity of Newfoundland possibly as early as the 1480's. The French and Spanish shortly there after with the Dutch slightly later, touching the continent at various locations on the eastern seaboard. For the early explorers, maps were not their concern other than navigation maps of the ocean upon which they sailed, reflecting landfall with a corresponding safe anchorage.

Samuel Blodgett's map which appeared in Boston in late 1755 (see Samuel Blodgett's Map of the Battle of Lake George by Edward Dodge in the Fall 2005 Newsletter) is unusual in several respects. It covers only two Phases of the battle and is printed so that the map is both horizontal and vertical in presentation (see Figure 1).

Blodgett's Map – Boston 1755.
Public domain.

Modern maps are orientated south to north. The left side of Blodgett's Boston published map showing Phase one of the battle or the "Bloody Morning Scout" is orientated north to south. The second Phase of the battle at the British main camp is orientated east to west as if you are looking down from the heights of French Mountain just to the east of the camp site. Blodgett was actually located in the center of the camp during the battle.

The top horizontal part of the map appears to have been drawn as an after thought to fill in a rather large open space on the map. It shows the route from New York City to Lake George and is orientated south to north yet depicted west to east on the map. Fort Lyman is shown as it existed. Fort William Henry is shown as apparently copied from Captain William Eyre's drawings of the fort as only the foundation had been laid by Colonel Ephraim Williams regiment prior to the battle on 8 September 1755. That the sketch of Fort William Henry came from Captain Eyre's architectural drawings is substantiated by the indicated linear footage of the curtains and bastions.

Finally, in the lower left corner of the map is a dedication to Governor William Shirley of Massachusetts (see Figure 1).

Blodgett in his Prospective-Plan; which accompanies his map, sets forth by number and description where the action took place as well as the location of opposing forces and camp equipment. Although much too lengthy to appear as an attachment to this article, this index covers thirty nine different points of reference. The final Phase of the battle or "Bloody Pond" is not shown on the map, in all probability as it was a running battle covering a distance of two to three miles with no knowledgeable cohesion. It is referred to in the Prospective-Plan by a rather extensive narration as is the route from New York to Lake George.

It is not known if the original document exists. Apparently Blodgett's original American produced map of the Battle of Lake George was long forgotten until a copy of it was found in a bound volume of the Boston-Gazette for 1755 which was presented to the Massachusetts Historical Society in 1883.

REFERENCES:

Blodgett's Prospective-Plan

An Introduction to The Study of Map Projections, J,A, steers, University of London Press Ltd, 14th ed. 1965

Principles of cartography, Edwin Raisz, McGraw Hill Book Co., 1962

Squanto and The Pilgrims -Native Intelligence, Charles C. Mann, Smithsonian, December 2005

The Illustrated Colombia Encyclopedia, Columbia University Press, 1969, Vol. 13

Holdings of The New York Public Library

Article appeared in Fall 2005 Fort George Advice

Blodgett's British Published Map of the Battle of Lake George

Both the American and London (British) versions of the map give a similar appearance (see Samuel Blodgett's Map of the Battle of Lake George by Edward Dodge, in the Fall 2005 newsletter). They are different in several ways. The London version has no horizontal element other than a descriptive narrative which reads "A Prospective View of the Battle fought near Lake George on the 8th of September 1755, between 2000 English with 250 Mohawks under the command of General Johnson & 2500 French & Indians under the command of General Dieskau in which the English were victorious, capturing the French General with a number of his men, killing 700 and putting the rest to flight".

There is no dedication in the lower left corner on the London version as there is on the American version. Map orientation is again different and placement depiction within the map is totally different (see Samuel Blodgett's American Published Map of the Battle of Lake George by Edward Dodge, in the Spring 2006 newsletter).

On the London published version, reading from left to right, the map of the route from New York to Lake George is shown first. By having the route moved from the horizontal position as shown on the American map to the vertical position brought a minor but obvious compression of the route on the London map. Here the route map is oriented from south to north in the style of modern maps. The plans for the forts are no longer at the north end of the map, but printed at the bottom extending to the reader's right under the depiction of the first phase of the battle. Fort William

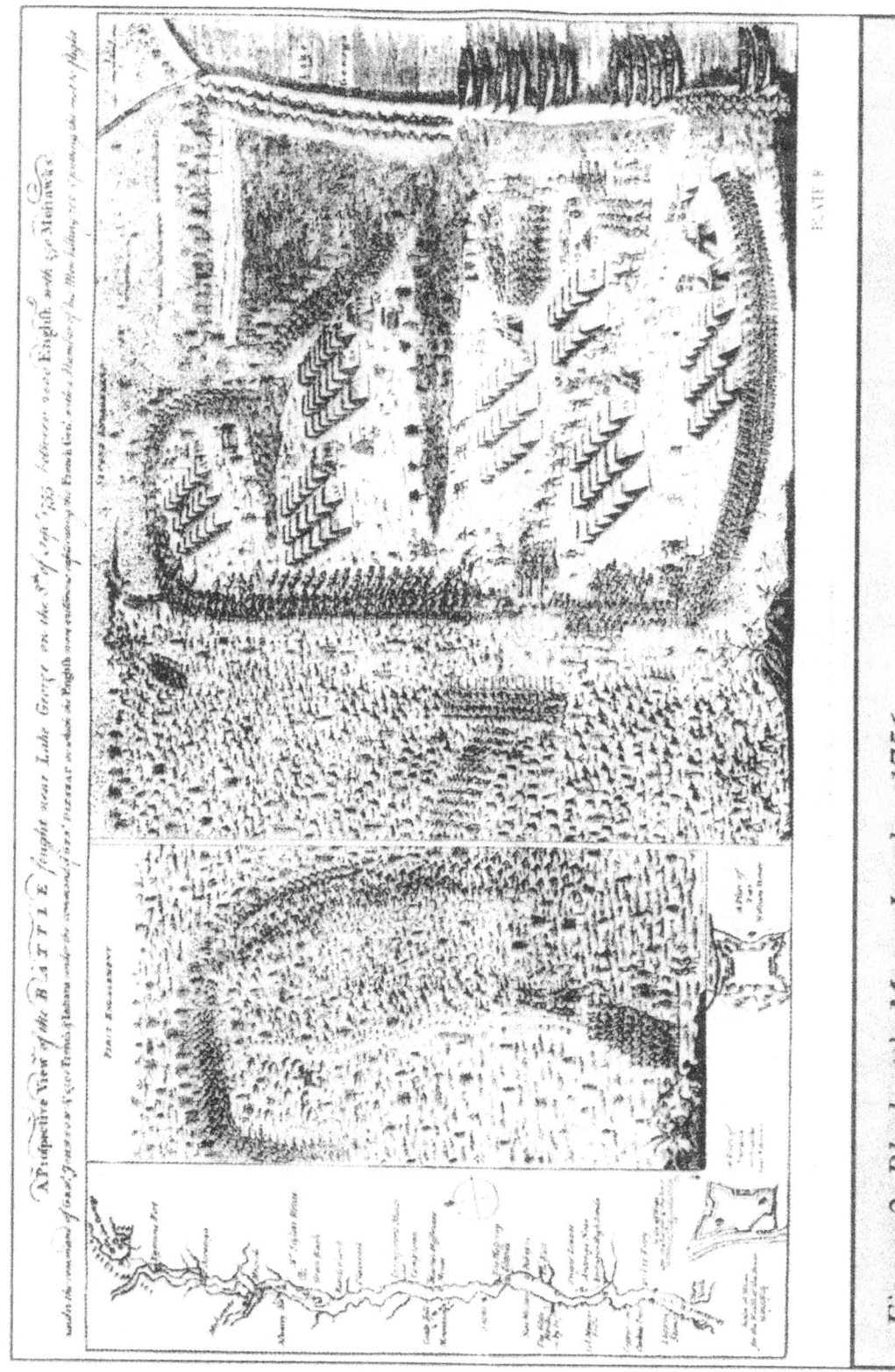

Figure 2: Blodgett's Map - London 1756

Samuel Blodgett's Map – London 1756.
Public domain.

Henry is depicted differently reflecting the shore of Lake George.

The map of Phase One of the Battle of Lake George or "The Bloody Morning Scout" is next as one reads from left to right. The map of "The Bloody Morning Scout" is oriented north to south. At the top it is labeled "First Engagement". Next to the reader's right is the map of the battle at the British main camp. That map is oriented east to west and labeled at the top "Second Engagement". As with the American version there is no map of the third phase of the Battle of Lake George. The map of the "Second Engagement " has been decompressed on the British version, making it easier to read as the spacing between objects is much more defined.

The London map was published by Thomas Jefferys with the Prospective-Plan reprinted for him at the corner of St. Martins Lane, London, selling for one shilling. There is circumstantial evidence that the London map was carved in resin which would explain the cleaner definition of the map.

The American version of the map was done by Thomas Johnston (Johnson), sculptor, coppersmith and Japanner. It is believed the version of the Prospective-Plan was printed by Richard Draper of Boston.

The Blodgett maps, particularly the London version, has been reprinted over the years in various publications referencing maps of early North America, warfare in North America and military art. The American version remained lost in time until 1883 when a copy was presented to the Massachusetts Historical Society. Various printings occurred in 1762, 1790, 1803, 1902, 1911, 1992, 2002, 2003 and 2005. Publication occurred in North America, England and Spain.

REFERENCES:

Blodgett's Prospective-Plan

An Introduction to The Study of Map Projections, J, A, Steers, University of London Press, Ltd, 14th ed., 1965

Principles of Cartography, Ewin Raisz, McGraw Hill Book Co., 1962

Squanto and the Pilgrims - Native Intelligence, Charles C, Mann, Smithsonian, December 2005

The Illustrated Columbia Encyclopedia, Columbia University Press. 1969, Vol. 13

Holding of the New York Public Library

Article appeared in Fall 2006 Fort George Advice

KING HENDRICK

King Hendrick was born Theyanoguin on 28 March 1692 in Westfield, Massachusetts. His mother was a Mohawk of the Bear Clan; his father was believed to be a Stockbridge Mohegan.

He was actually the second King Hendrick. The first was Tejonihokarawa, born in 1660, a member of the Wolf Clan. A painting of him done in 1710 is often taken as a painting of the young King Hendrick, although his clan totem is clearly visible in the painting. H died in 1735.

Theyahoguin was also known as Henry or Henry Peters with the French calling him "Tete Blanche" or White Hair. He stood about six feet tall, was slender with three tattoo lines across his forehead. He had a pronounced scar on his left cheek and wore his hair in a scalp lock.

His life as we know it, with one exception, covers only 15 years. As a child he moved with his mother to Canajoharie a major Iroquois village in New York. It was there in all probability his brother Abraham was born.

In 1722 Hendrick traveled to Massachusetts in an attempt to make peace between the whites and Abanaki. This attempt was foiled by the whites who felt he was on a recruiting trip. Recruiting for whom?

Hendrick visited England in 1740 where King George presented him with a suit of clothes and a cocked hat. An engraving was done of him wearing the suit and hat, which he wore to the Albany Conference of 1754.

King Hendrick, Mohawk Chief.
Courtesy of Williams College Library, Archives and Special Collections,
Williamstown, Massachusetts.

Hendrick went to Boston in 1744 to make treaties and pacts with the British. He attended Governor Clinton's conference in Albany. The conference accomplished nothing except bruised feelings.

In 1746 Hendrick and other Mohawks traveled to Montreal where they received gifts of good will from the French. On the way home they attacked a party of French on Isle LaMotte on the Richelieu River. Hendrick had made the French "Most Wanted List."

During the spring of 1747 Sir William Johnson, Hendrick's friend, had Hendrick and the Mohawks savage the French in the Lake Ontario-Oswego River area. In 1748 Hendrick and other Mohawk chiefs were invited to talk again with the French. Probably Abbe Francois Picquet of La Presentation at what is today Ogdensburg, New York, extended the invitation. There is no evidence that it was accepted.

Sir William Johnson had a red coat/suit which Hendrick admired. Johnson gave Hendrick the coat/suit after Hendrick explained he had dreamed about it. Johnson told Hendrick that he had dreamed that Hendrick had given him five hundred acres of Iroquois land. Hendrick gave him the land but told him "I will never deam with you again."

IN 1750 Hendrick tried, with Johnson, to get the western (Ohio Territory) tribes to align with the British, but they were unsuccessful.

Hendrick was present at the Albany Conference of 1754 which ended on a negative note although treaties between the British and most of the Iroquois were signed. He attended another conference in Philadelphia on January 1, 1755, arriving back in Canajoharie in May.

August 1755 found Hendrick, his son Paulus and his brother Abraham at Lake George. It was here that he delivered his famous "One stick will break, several will not" speech; adding that the force Johnson was going to send out to meet the approaching French "Were too few to fight and too many to die". Johnson sent a larger force that was ambushed and Hendrick who had gone with the force was killed. Or was he? Paulus and Abraham were not injured.

There are several stories about his death. One says his horse was killed and he was pinned under it and bayoneted by a French soldier. Another states that an Indian boy shot at him with a bow and hit his horse with the preceding outcome. Yet another story relates that he was pulled off his horse and butchered by Indian women travelling with the French.

Are any of the stories correct? His body was never found. Was he captured? Tortured and killed elsewhere? Was his body so badly mutilated that it could not be identified? Questions with no answer.

Hendrick was fluent in language, a diplomat, orator and war chief. He was loyal to his clan, tribe, nation and the British. That is a legacy a man could wish for and be proud of.

REFERENCES:

Turtles, Wolves and Bears, A Mohawk Family History by Barbara J. Silversten, Heritage Books, Bowie, Maryland, 1996

Theyanoquin, In Northeastern Indian Lives, 1632-1816, edited by Robert S. Grumet, pp 208-226, The University of Massachusetts Press, Amherst, Mass.

Dean R. Snow, Chair – Department of Anthropology, Pennsylvania State University, University Park, Penn.

Relief is Greatly Wanted, The Battle of Fort William Henry, Edward Dodge, Heritage Books, Bowie, Maryland, 1997

The Summer Paradise in History, Warwick S. Carpenter, 1914, The Delaware & Hudson Company, Albany, NY

The Invasion Within, James Axtell, 1985, Oxford University Press. New York, NY

Montcalm & Wolfe, 1884, Francis Parkman

The Illustrated Columbian Encyclopedia, Vol. 10, pg 2844

EPHRAIM WILLIAMS

In one of history's mysteries two leaders would die within minutes of each other. They were King Hendrick the Mohawk chief and Colonel Ephraim Williams, Massachusetts Militia. In both instances we really know only of the last 15 years of their lives.

There are at least three different stories relative to King Hendrick's death and two relative to Colonel Williams'. One story states that the Colonel died from a head shot fired during the first volley fired by the French and Indians. The second story states his horse was killed during the first volley and after getting free of his mount he climbed a boulder and was rallying his men when he was killed instantly by a bullet in the head.

The ambush took place a short distance from Lake George, New York on the military road between Fort Lyman (Fort Edward) and the site of the future Fort William Henry. The French and Indian force under the command of Baron Dieskau had originally targeted Fort Lyman. When it was learned that Fort Lyman was heavily endowed with cannon the French Indians refused to attack the fort. Dieskau determined that the forces under Sir William Johnson at Lake George would be ripe for attack. Accordingly, Dieskau left the Fort Lyman area early on 8 September 1755 and headed up the military road. His scouts reported the column that Johnson was sending to relieve Fort Lyman and an ambush as set. The ambush became known to history as "The Bloody Morning Scout" and the subsequent battle at Johnson's position on Lake George as "The Battle of Lake George."

Ephraim Williams, Jr. was born to Ephraim, Sr. and Elizabeth Jackson Williams at Newton, Massachusetts on 24 February 1715. His brother Thomas was born at Newton on 1 April 1718. He had other siblings, but they were of different parentage.

His father was a landholder, farmer and surveyor as well as one of the original English families who helped survey and lay out Stockbridge, Massachusetts, which became known as "The Home of The Praying Indians."

The individual Indians became known as the "Stockbridge Mohigans" and would become known for their tutelage, friendship and allegiance to Major Robert Rogers of Ranger fame.

Ephraim apparently had the equivalent of a grade school education and did become a trained surveyor. Family lore states he became a sailor at a young age and made multiple trips to Europe, but there is no documented evidence to support this.

Ephraim joined the Massachusetts Militia during King George's War (1740 to 1748) apparently as a private. He was a member of Governor Shirley's expedition to Canada which never got to Canada. His service was on the northern frontier. He acquitted himself well and was promoted to permanent Captain with a brevet to Major on or about 29 July 1745. He reverted to his permanent grade of Captain on 7 February 1746.

In 1750 the government of Massachusetts granted Ephraim 200 acres of land between Williamstown and Adams, Massachusetts. In 1751 he erected Fort Massachusetts on his land and on 18 August 1751 wrote a proposition for the picketing of the fort. The fort was located in the extreme northwest corner of Massachusetts in juxtaposition to the Hoosic River and what would become known as the Appalachian Trail, in part would be known a the Mohawk Trail. The fort was palisaded, 120 feet long and 80 feet wide. The walls were 16 feet high with the palisades being 6 to 14 inches in width, the larger being weight-bearing. The gate was in the north wall and the parade was on the interior north side of the fort. The well was in the northeast corner, with the watch box being on the northwest corner. Individual quarters and/or barracks were up against the east, west and south walls with the palisades forming the back wall of the quarters. The roofs of the quarters sloped upward to meet the palisades creating a salt box type of construction. This was the first of several forts of the same basic construction that he was involved in building on the northern frontier.

While at Fort Massachusetts Williams went out during one attack to bring in some settlers who were trying to get to the fort and hotly pursued by enemy forces. He got them safely into the fort and just barely escaped being cut off. On another occasion while he was away the fort was attacked and razed and the captives taken to Canada. He was promoted to Major on 7 June 1753.

He was during this time commander of all forts west of the Connecticut River and functionally the northern frontier. He was in Boston on 10 January 1755 and received recruiting orders from Governor Shirley on 11 February 1755. On or about 28 April 1755 he was promoted to permanent full Colonel. On 11 May 1755 he received marching orders from Governor Shirley as he had raised a regiment. In keeping with his promotion he ordered and paid for a new set of Regimentals from Henry Loughton & Co. On or about 11 June 1755. Governor Shirley gave him a commission to administer oaths of office to officers and others on 11 June 1755 and he and his regiment left for Albany, New York to join the forces of Sir William Johnson on the New York frontier.

Williams and the regiment arrived in Albany on or about 22 July 1755 after a cross-country march of roughly forty days. According to legend and supported by dated material, Williams made his will on the 22nd of July as he had a premonition of his death. He sent his will and a letter to his cousin Israel Williams whom he designated as his executor/administrator.

By 27 July 1755 he was a Fort Edward with the regiment. During August 1755 he wrote Israel another three times. 28 August 1755 saw him and the regiment at Lake George. Sir William Johnson directed Williams and his men to clear the promontory overlooking the lake and to lay out the lines of the fortification. Fort William Henry was started by Williams and would be finished by Captain William Eyre of the 44th Foot.

On 8 September 1755 Colonel Williams and King Hendrick were killed during the early morning hors as a result of an ambush by French and Indian forces.

By evening of 8 September 1755 the battle was over and the colonists took count. A casualty list for Colonel williams' regiment was done and he was listed as Killed in Action. On that date a Thomas Clark wrote Isreal Willliams a letter advising him of Ephraim's death.

Was Thomas Clark, Ephraim's younger brother Thomas, a physician assigned to the regiment? He was involved in the ambush, survived and spent the day following his calling. He would later be commissioned s Lieutenant Colonel in 1756 and saw service at Lake George.

Under terms of his will the majority of Ephraim's property and funds were to accumulate for thirty years and then be used to start a free school. The school was later converted to a college and on 2 September 1795 saw it's first graduation of four students. The college now known as Williams College, graduates an average of 450 to 500 students a year.

Colonel Williams never married. Historically he is described as a large, fleshy man. As a point of reference, Sylvia Brown, College Archivist, Williams College, indicated the college has a coat that belonged to Ephraim's brother Thomas. A woman of normal size of our time cannot put it on as it is so small. Pure supposition would indicate that Ephraim, because of his military and fort building activities, probably was 5'6" to 5'8" in height, broad shouldered and barrel chested. There are no known portraits of him.

In 1854 the alumni of Williams College created a monument to Colonel Williams at the spot where he was killed. The monument is an engraved boulder surrounded by a protective rail fence, located off Route 9 North just south of Lake George, New York. The Colonel's remains have since been moved to a permanent site on the college grounds.

REFERENCES

Sylvia Brown, College Archivist, Williams College, Williamston, Massachusetts

The Massachusetts Historical Society, Abigail Online, Israel Williams Papers

Appleton's Cyclopedia of American Biography

The Illustrated Columbian Encyclopedia, pg. 6691, Vol. 22

Relief is Greatly Wanted, Heritage Books, Bowie, MD, Edward J. Dodge

The Summer Paradise in History, Warwick S. Carpenter

Origins in Williamstown, Arthur L. Perry

The empire State Scrap Book, Ernest E. Bisbee

Fort William Henry – A History, Lake George, New York, Stanley M. Gifford

Major General Seth Pomeroy

Seth was born May 20 1706, in Northampton, Massachusetts, the seventh child and fifth son of the Honorable Major Ebenezer Pomeroy and his second wife, Sarah King. Their other children were: Sarah (1693-abt. 1693), John (1695-1736), Ebenezer (169-1774), Sarah (1700-1777), Simeon (1702-1725), Josiah (1703-1789), Daniel (1709-1755), and Thankful (1713-1790).

Seth was home educated, as was the practice at the time. The greater part of his education was toward learning the trade and principles that he was to follow and live by as a mature man. His father Ebenezer, his paternal grandfather Medad, and his paternal great-grandfather Eltweed were blacksmiths and gunsmiths by trade. Seth developed a reputation of being a religious man, strong in his convictions, kindly, honest, friendly, hardworking and not tolerate of fools. There are no known pictures (paintings) of him. However, one account has him being a man of great strength and ability, tall, spare and erect.

At the age of 26, Seth married Mary Hunt on December 14, 1732. For years, the couple lived on the family homestead in Northampton where their nine children were born. Seth (11733-1770), Quartus (1735-1803), Medad (1736-18129), Lempel (1738-1819, Martha (1740-1803), Mary (1742-1762), Sarah (1744-1808), an unnamed stillborn infant (1747) and Asahel (1749-1833). At Northampton he was involved in a number of land transactions and operated a gunshop.

The first known record of Seth's military service has him appointed, on January 23, 1743, Captain of the Third Company of snowshoe men being raised under Colonel John Stoddard. The same year (1743), he was a Militia Captain in the service of the Province of Massachusetts Bay. In 1745, he was commissioned a Major and served as the Captain of the 3rd Company of Colonel Joseph Dwight's Regiment during the expedition against Cape Breton during which he fought at the siege and capture of Louisbourg, Nova Scotia, Canada. In 1746, he was placed in command of the frontier after the April attacks upon Charleston and Keene, New Hampshire. On June 15, 1746, he was commissioned a Captain of Colonel Joseph Dwight's Regiment in which he was actively engaged through 1747. During some of that period of time, he commanded Fort Massachusetts. In July and August 1748, he was invoved in scouting expeditions against rhe French and their allies.

In 1755 during Massachusetts Governor Shirley's planned/expedition to attack Crown Point on Lake Champlain, Seth served as a Lieutenant Colonel in Colonel Ephraim Williams Regiment. (See Ephraim Williams by Edward Dodge in the Fall 2004 newsletter). Following the Battle of Lake George (see The Battle of Lake George - 8 September 1755 by Edward Dodge in the Spring 2005 newsletter), Seth reported himself as "being the only Field Officer in Colo. Ephraim Williams Regiment supposed to be now living". Seth lost his younger brother Daniel on that day worthy of memory (September 8. 1755) to enemy fire. During the remainder of the French and Indian War (The Seven Year War), Seth saw service in 1757 in an aborted attempt to relieve the French siege upon Fort William Henry (the Fort surrendered before the relief force could arrive), in 1759 and 1760 at frontier forts and from April 5th through June 20, 1760 (22 weeks) during the West Hoosuck expedition.

Seth was a delegate to the Massachusetts Provincial Congress in 1774 and 1775. On February 9, 1774, that Congress appointed him a "General Officer" among others, all without more definite rank. On October 27, 1774, the same Congress appointed him third in command of the Massachusetts forces. Jedidiah Preble was first in command and Artemus Ward, second. Ward was commissioned commander-in-chief in May 1775, and John Thomas made Lieutenant General, while Seth Pomeroy was commissioned a Major General, Preble having retired. The Massachusetts House of Representatives fix the pay of its General Officers on January 25, 1776, naming Seth as a Major General. He drew pay at that rank on that date for two months and nine days.

While briefly at home in Northampton for a few days rest in June 1775 (a brief 24 hours), Seth was summoned by General Isreal Putnam to repair to Boston. Starting on June 16th, he rode all night and reached the site of the Battle of Bunker Hill (The Battle of Breeds Hill) at two o'clock in the afternoon. Here he fought through the battle as a private soldier, having refused repeated urgent offers of general command. Here, Seth, the old warrior advanced into the trench and took charge of the Connecticut troops. With a gun of his own making, which he had carried thirty years before at the siege of Louisbourg, he directed the fire of his men during those two hours of struggle for the birth of American liberties.

On June 16, 1775, the Continental Congress authorized the appointment of the first eight Brigadier Generals of the American Continental Army. On June 22nd Congress appointed those officers, designating them by number. Seth Pomeroy was the first and senior. Nothwithstanding that, his appointment was short lived. General John Thomas was appointed first Brigadier General in the Army of the United Colonies by an Act of Congress on July 19, 1775. The US Military Academy at West Point today memorializes the first Brigadier Generals of the Army and Seth's name heads the list.

In the 18th Century, it was customary for General Officers of the various Provincial/State Militia to also hold Commissions of lesser rank and command Regiments. Massachusetts was no exception to this practice. While serving at the head of the Massachusetts Militia, Seth Pomeroy also commanded his own Militia Regiment from Hampshire County. On January 31, 1776, the Massachusetts House of Representatives appointed him Colonel of the Second Hampshire County Regiment of Militia. When he entered service in 1777 as head of the Massachusetts Militia destined to join General Washington in New Jersey, his regiment went with him.

Seth died of pleurisy at Peekskill, Westchester County, New York on February 19, 1777 while he was traveling with his Massachusetts troops to meet General Washington. Seth was buried in the Old Van Cortlandville Cemetery in Van Cortlandville, Westchester County, New York in an unmarked grave.

Seth Pomeroy lived and died as a man of strong convictions, both to his family, his God and his Country.

SOURCES:

The MEADOW City's Quarter Millennial Book 1657-1904, Northampton, Massachusetts, 1904

The Journals and Papers of Seth Pomeroy a sometime General in the COLONIAL Service, Ed. Louis Effingham de Forest, A.M. J.D., The Tuttle, Morehouse & Taylor Company, Salem, Massachusetts, 1912

The History and Genealogy of the Pomeroy Family Collateral Lines in Family Groups, by Albert A. Pomeroy. Reprinted Higginson Book Company, Salem, Massachusetts, 1912

Hampden County, MA Deed Records Index, Grantees Index A-Z, FHC Film #844472, Salt Lake City, Utah

American Firearms Makers, by A. Merwyn Carey, New York; Thomas Y. Crowell Company, 1953

Massachusetts Soldiers in the Colonial Wars, New England Historic Genealogical Society online [http.//wwwnewenglandancestors.org/research/]

Peekskill in the American Revolution, by Emma L. Pattersom, The Friendly Town Association, Inc., Peekskill, NY, 1944

Article appeared in Fall 2007 Fort George Advice

A SECOND LOOK

Throughout history novelists and historians have used various wars or battles as a focus point for their writing. I originally wrote about "The Battle of Lake George" in 2005. (See The Battle of Lake George - 8 September 1755, in the Spring 2005 newsletter).

Numerous sources were used, however Samuel Blodgett's "Prospective Plan of The Battle of Lake George" was the primary source of information. In the three years since then more information about the battle has come to light. Lieutenant Colonel Seth Pomeroy's papers and journal have provided a great deal of information from a real time aspect.

The Colonel's journal indicates that the troops arrived at Lake George at 4PM on Thursday, 28 August 1755, with a strength of 2000 men inclusive of Colonel Ephraim Williams Regiment of which Colonel Pomeroy was a member, serving as a company commander. (See Ephraim Williams, Fall 2004 newsletter).

His journal entry of Monday, 8September states that in view of information they had received on the 7th of a large body of men traveling south near Wood Creek (east of the lake separated from the lake by a mountain) approximately 1200 men, 200of them Indians using the military road moved south toward The Great Carrying Place, aka Fort Edward, Fort Lyman, Lydius Trading House or Fort Lydius. See (Ephraim Williams, Fall 2004 newsletter) and (King Hendrick, Spring 2004 newsletter).

The troops had gone about three miles when they were ambushed and fired upon by the French and Indians, The front of the column fought bravely, but many of the men in the rear fled. Others remained and fought a very hamsome retreat by firing, retreating a little and then rising and giving the enemy a brisk fire. This continued until they were within ¾ of mile of the camp. At that point the troops gave a last volley killing great numbers of them - seen to drop like pigeons and stopping the French momentarily. Some 300 men from the camp were sent out under the command of a Lieutenant Colonel Cole to assist the retreating troops.

The French soon drove on with courage with the French Regulars about six abreast and came within 20 rods of the camp. The French went to the left of the cqamp as you faced them and took shelter behind trees, logs and other places where they could hide.

The troops in the camp drew 3 or 4 field pieces into a relatively safe position with other pieces toward the west part of the camp. Colonel Pomeroy placed himself with his troops and some of Colonel Ruggles men in a fire line toward the west of the camp. The battle started in ernest between 11 and 12 o'clock and continued until 5 PM.

The French General who was wounded and about 30 of his men were taken prisoner. There were a considerable number of French dead and the General advised them that th greater part of his officers were killed and his army or force broken. He did not know the number of dead, but byody count and prisoners put the number at not less then 4 or 500 men.

On Tuesday, 9 September they buried the dead in camp and reinforced their position. No number was given for the dead in camp.

The colonel's journal entry for Wednesday, 10 September, indicated that it was fair and hot. He went out with 400 men to bury the dead killed on the 8th. Apparently at the ambush site and along the road. They buried 136 men in what the colonel termed "A most melancholy piece of business". The body of Colonel Williams was found and buried. There is no mention of the body of King Hendrick being found. Colonel Pomeroy states that after Hendrick's horse was shot, that Hendrick was on foot heading toward the troop column when he was attacked by a group of women and boys from the French Indian allies who stabbed him in the back with a spear or bayonet , killing him and taking his scalp.

The colonel also learned that his brother Daniel-born in 1709 was killed and scalped. Daniel was a Lieutenant in Captain Elisha Hawley's company, Williams Regiment. The detail with Colonel Pomeroy collected 2 0r 3 wagons full of French stores, guns, blankets, hatches and brought them into the camp.

Journal entry for Thursday, 11 September, indicates more men sent out to pick up what stores could be found and to bury any other bodies found. They buried four more od their own and found a great number of French bodies both hidden and buried. Evidence was that the French left in fright.

On 9 September, Colonel Pomeroy wrote to Colonel Israel Williams to advise him of the events of the 8th. As of the 9th Colonel Pomeroy was the only surviving Field Grade officer from Williams regiment.

A point of contrast is that in his journal Colonel Pomeroy indicates that 300 men under the command of Colonel (Lieutenant Colonel) Cole of Rhode Island left the camp to reinforce the retreating column. In his letter of 9 September he indicated that Colonel Cole was in charge of the rear of the column. Other sources indicate Lieutenant Colonel Whiting of the Connecticut regiment was Colonel Williams number two and in charge of the rear section of the column. Battlefield confusion.

The letter indicates that not only was the French General Baron de Dieskau captured, but his ADC (Aide de Camp) was captured as well. Through questioning it was learned that the French had 4000 men at Crown Point (Fort St. Frederic) and the immediate French column had 1800. Captured papers had plans for all of General Johnson's operations, maps of Fort Edward and a lay out of the camp at Lake George. The colonel ends the letter with the hope that the prayers of God's people will sustain and support them in their attack on Crown Point.

On 11 September, the colonel wrote his sister-in-law Rachel Pomeroy, wife of Daniel, to inform her of Daniel's death. He was very straight forward, telling her that he had been killed by a fatal shot through the middle of his head. He did not tell her that he had been scalped. He wrote further that they had buried 136 and had since buried 20 more. For some reason he told her of their bringing in a wounded man who hah been setting where they found him (he didn't say where) since 8 September with the back of his skull cut off and his brain exposed. He did not think the man would live. He closed with the hope that God's blessings would be with her and the children and indicated the family would help in all ways possible.

Williams regiment sustained 50 dead, 21 wounded and no missing as a result of 8 September. They brought in 27 French prisoners of whom 20 were wounded.

Their equipment loss was exceedingly low: 12 blanket haversacks, 11 pumplines (probably trumplines-used to distribute weight of haversack load), 5 hatches, 2 waistcoats, 2 coats, 2 shoes, 2 hats, 2 caps, 1 knife and no listed fire arms.

Lieutenant Colonel Pomeroy was promoted to full Colonel and took command of the regiment on 9 September. 1755.

REFERENCE;

Journals and Papers of Seth Pomeroy as published
By The Society of Colonial Wars in the State of New York, 1926

TIMOTHY RUGGLES

Jurist, politician, military officer, strong willed, determined. Such was this man born 20 October 1711 who died 4 August 1795. This combination created a life that was full of hard work, affluence, respect, extreme hatred and personal pain we cannot even begin to imagine.

The Ruggles family came from Eure in Normandy, France where the name was DeRuggele. The family removed to England after 1066, where the spelling of the name became Ruggeley. In England, the family swerved as a Fellow at Clare College, one of the founders of the Virginia Company, a playwright of such renown as to receive the appreciation of James I - King of England, served as High Sheriff of Suffolk and Representative of the county of Essex in Parliament

The family removed to Roxbury, Massachusetts in 1637. Timothy was born as the first child of Reverent Timothy and Mary (White) Ruggles on 23 October 1711. In North America the spelling of the family became Ruggles. He was one of twelve children. He was schooled and had his schooling directed by his father. It was such that he passed the entrance examination to Harvard from where he graduated in 1732 at the age of twenty one.

He chose law as his profession and practiced it in Rochester and Sandwich, Massachusetts until he settled in Hardwich, Massachusetts in 1753. He married a widow Mrs Bathsheba Newcomb in 1735. By 1753 his practice had grown and he had accumulated a liberal fortune. He had an extensive estate with a stable of thirty horses, a extremely large deer park and a goodly size pack of hounds. Visitors were lavishly entertained.

He was appointed Judge of the Court of Common Pleas in 1756 and from 1762 to the start of the Revolution, he was Chief Justice of that court and according to the Judicial History of Massachusetts he was faithful, able and incorruptible.

He was further described as over six feet tall, well proportioned with stalwart shoulders and a noble head. President Adams stated in 1759, " That Ruggles grandeur was in the quickness of his apprehension, the steadiness of his attention, the boldness and strength of his thoughts and expressions, his strict honor, conscious superiority and contempt of meanness. People approached him with dread and terror. He is proud and lordly and it is easy and natural in him."

In 1755 Ruggles was a Colonel in charge of a Regiment under Sir William Johnson at Lake George, New York. This was Governor Shirley's planned attack against the French at St. Frederick (Crown Point) on Lake Champlain. Governor Shirley was the Governor of Massachusetts and considered by the other provincial governors as their leader.

Ruggles Regiment was sited and picketed as the third row back from the southern boundary of the camp, on the west side of the camp next to the swamp which separated the camp from what would be the site of Fort William Henry. During the attack by the French and Indian forces in what is known as The Battle of Lake George (see Fort George Advice - Fall 2006). his regiment fought under the command of Lieutenant Colonel Seth Pomeroy on the west side of the line. I have found no information as to Ruggles location during the battle. (See Fort George Advice - Spring 2008 - A Second Look at the Battle of Lake George, September 1755).

In 1758, Ruggles was in command of the Third Division - Provincial Troops under General Abercombie during the General's unsuccessful and costly attack against Fort Carillon or Ticonderoga as the British Called it. In 1759 he was again involved in an attack on Carillon under the very able General Amherst. There are records showing that Brigadier General Ruggles 2nd Battalion was present. On 15 July 1759 he had five boats assigned to him for his family. No further information is given. On 17 July 1759 he was Colonel of The Day. His 2nd Battalion was drawn up two deep from the right shore of Lake George, while his 1st Battalion was in the right column with the artillery. General Amherst defeated the French.

In 1762 Ruggles was chosen Speaker of The House of Representatives. In 1763 as a reward for his military service he was appointed "Surveyor General of the King's Forests". This was in the nature of a sinecure, carrying considerable honor, dignity and 3000 pounds annually.

In October 1765 he was chosen as one of the delegates from Massachusetts to The First Colonial Congress and The Stamp Act Congress. Both were really the same meeting. He was chosen President of The Congress.

The congress adopted memorials to the two Houses of Parliament and a petition to the King praying for more humane policy and justice to the subjects of America while professing loyalty to the King.

Ruggles disagreed as he felt the documents were the seeds of revolution. For this he was censured and reprimanded by Massachusetts. He did not change his opinion. His wife and several of his children did not share his feelings and he incurred the hatred of his neighbors and friends. The Crown appointed him one of the "Mandamus Council". His younger brother warned him if he took the oath and became a member of the "Council" he would never be permitted to return alive. The "Council" was for all intent and purpose the "Royal Court in North America".

He took the oath. His devotion to duty as he saw it, cost him everything. Family members would serve with General Washington. His wife and eldest son stayed loyal to America. From the time he took the oath until his death, he never saw the face of his wife again.

As a "Mandamus Councilor" in Boston he raised a Battalion of two hundred plus, composed of merchants known as "The Gentlemen Volunteers" or "Loyal American Associates". The Battalion left Boston when the British forces left. It has been claimed that he and his unit went to Long Island, but there is no real evidence to support the claim.

In the Act passed in September 1778 by Congress, forbidding the return of loyalist refugees under penalty of death, Ruggles is number four on the list.

It is not precisely known when Ruggles left for Nova Scotia. Records put him in Annapolis, Nova Scotia in 1783. He filed an application for a grant of lands and received ten thousand acres in the town of Wilmot, Annapolis County, Nov Scotia. He rebuilt from scratch a large roomy house, orchards and successfully raised a multitude of various types of trees and shrubbery. His wife died in 1787,. His two younger sons lived with their father. His daughters stayed with patriotic husbands in Massachusetts

On 15 February 1746, Ruggles and his wife were delivered of a daughter "Bathsheba" named after her mother. In 1766 Ruggles arranged for the marriage of Bathsheba to Joshua Spooner a well to do farmer from Brookfield, Massachusetts. The couple had four children, a daughter and three sons. By 1778 Bathseba had developed an aversion to her husband. It was alledged that he was abusive. She met and is said to have taken as a lover a young continental soldier. When she found she was with child she told her lover to kill her husband. He could not or would not do it so she recruited two escaped British prisoners of war and they beat her husband to death.

Drink. Loose talk and personal property of the deceased in possession of the British prisoners of war brought about a trial. At the trial it became evident that Bathseba was involved. It was attempted to prove she had a disorder mind, but it was not believed and all parties were found guilty and sentenced to be hanged including her reluctant lover. She tried for postponement and even though the pregnancy was confirm, the finding was not accepted and she and the three men were all hung on 1 July 1778. It was determine after her death that she had in fact been carrying a healthy five month old male fetus.

Ruggles must be admired for staying with what he felt and saw as being right. Was it worth it? He lost a country, wife, most of his family, all his friends and even worse a daughter in an arranged marriage that resulted in her being convicted of a felony that resulted in her death and the death of her unborn son. You draw your own conclusions.

Ruggles died on 4 August 1795 at the age of 84. He is buried in an unmarked grave eastward of the Chancel of the church in Wilmot, Nova Scotia.

REFERENCES:

General Timothy Ruggles by Henry Stoddard Ruggles, Wakefield, Mass. Privately printed – 1897

Page 280 – Papers of Sir William Johnson – Minutes of Council of War, 18 August 1755

Once Again a Return to The Battlefield

The camp site is basically gone. The wounded in hospital and the dead put to rest. The air is clear with only a trace of the wild onion that normally grows here. Pieces of clothing, tarp and badly damaged but identifiable equipment can still be found. Tracks from wagons, horses, oxen and hundreds of humans mar the ground. In the near distance the noise of the men building Fort William Henry can be heard as can an occasional musket shot further away - hopefully fresh game or fowl.

We know or have a fairly good idea of the men who fought here, particularly the leaders. We know what happened to General Johnson, Colonel E. Williams, King Hendrick, Colonel Ruggles and Colonel Pomeroy. The others? Mere shadows in the mists of time? The mists have lifted, not completely, so those shadows are no longer shadows, but faint glimmerings.

General Phineas Lyman conducted the defense of the camp. He was born in 1716 in Durham, Connecticut. Attended Yale University, graduated in 1738 and taught there until 1741. He then studied law and began practice in Suffield, Connecticut. By 1749 he was a member of the Upper Chamber of the Connecticut legislature. In March 1755 he was appointed Major General and made Commander in Chief of the Connecticut Militia force of 1000 men which participated in the expedition against Crown Point (Fort St. Frederic). He was in command of Fort Edward (Fort Lyman) for a time in 1755.

In 1758 he commanded the Connecticut troops in General Abercrombie's disastrous attack on Fort Carillon (Fort Ticonderoga). The following year (1759) he was with General Amherst when Carillon and Fort St. Frederic were taken. In 1760 he was with the expedition to Fort Oswego and Montreal. In 1762 he was in the invasion of Havana, Cuba. He went to England in 1763, staying until 1772, obtaining a grant of land in West Florida near Natchez. He led a group of settlers to that region in 1773. He passed away in 1774 and was buried at his home on the Blackwater River near Natchez, Mississippi.

Colonel Moses Titcomb was born on 8 July 1707 in Newbury,

Massachusetts, the 8th son of William and Ann Titcomb. He fought at the Battle of Louisbourg as a Major in charge of a battery of 40 - 2 pound cannon which were so effective it was noted in dispatch. He was apparently at that time with the Massachusetts 5th Regiment. On 8 September 1755 he was at Lake George with the regiment he commanded, the 3rd Massachusetts. During the battle, his regiment was on the right of the line. He along with a Lieutenant Barton had taken position behind a large tree slightly in front of the line from where he could get a good view of the enemy and direct his troops fire. History states that at approximately 4PM, he and the lieutenant were both killed by Indian snipers (it is alleged) located about 80 yards away in the swamp between his position and what would eventually become the location of Fort William Henry.

Lieutenant Colonel Jonathan Bagley of Colonel Titcomb's regiment took command of the regiment upon Titcomb's death. Jonathan Bagley was born on 23 March 1717 in Amesbury, Massachusetts. He was the sixth child of Orlando and Dorothy Harvey Bagley. There were twelve children in the family, seven boys and five girls. His father was a large land owner, teacher, selectman and trial justice. Jonathan married Dorothy Wells the daughter of Amesbury's minister on December 9, 1736. They had seven boys and four girls. Jonathan was husbandman living at the Amesbury Ferry in a house build by his uncle Timothy Currier. In 1741 he built a fifty foot wide wharf which protruded from the bank to the channel of the Pow Wow River at its junction with the Merrimac River. In 1750 Jonathan purchased another house, land and an adjacent ferry. By 1750 he owned as well, several vessels; one a sloop that he built, captained and sailed to the West Indies.

When his father died, he inherited nine acres and for many years he purchased and developed lands in Massachusetts, New Hampshire and Maine, earning the title of Esquire. In 1751 he built a third home in Salisbury, next to Amesbury. In 1761 he purchased Stephen Emery's grist mill. For service in the Seven Years War (The French and Indian War) he was granted land in Cumberland County, Maine. In 1768 he and Colonel Moses Little settled families on this land. He became a pioneer in Maine's lumber business, owning his own saw mill. He also had a lime kilm in Massachusetts on the Merrimac River. He became so wealthy that between the years 1759 and 1769 he paid more taxes then anyone else in Amesbury East Parish.

He was elected to the Massachusetts House of Representatives in 1743 serving eleven terms. He served in a variety of public offices. His military career started in 1740 when he swerved as a company clerk. In 1743 he was promoted too Lieutenant of Militia. In 1746 he was Captain of the 5th Company. 5th Massachusetts Regiment under Colonel Robert Hale and was involved in the 1745 assault on Louisbourg, Nova Scotia. In 1755 at the age of thirty-eight. He served as a Lieutenant Colonel and was promoted to Colonel the same year. He was commissioned a Colonel in 1756 to 1762, 1767 to 1769, 1773 to 1774. He left the army at age fifty-seven and retired to civilian life. He was raised to Master Mason on 18 October 1770 st. John's Lodge Number One, Portsmouth, New Hampshire. He passed away at the age of sixty-three at home on December 28, 1780.

Colonel William Cockcroft of the New York Regiment is truly a shadow. He was commissioned by the Governor of New York on 11 June 1755 to command the New York Regiment in the expedition against Crown Point (Fort St. Frederic). His Regimental Surgeon was Doctor Peter Middleton. The Colonel and doctor were commissioned the same date the regiment was raised on 21 August 1755 in Poughkeesie, New York.

Colonel Christopher Harris of Rhode island was appointed in March 1755. He had four companies of 100 men each in the regiment. Lieutenant Colonel Cole who was in charge of the 350 man relief element that went to help the survivors of Colonel Ephraim Williams column, was a member of that regiment. Lieutenant Colonel Cole was promoted to full Colonel and given command of a Rhode Island Regiment by order of General William Augustus Howe at the request of Captain James Montresor, Chief Engineer in America, British Army.

Lieutenant Colonel Nathan Whiting was born 4 May 1724 in Windham, Connecticut. His parents died when he as a child; so his uncle The Reverent Thomas Clasp and his wife raised him. Whiting graduated from Yale in 1743. His uncle, Reverent Thomas Clasp was President of the University at the time.

Whiting joined the New England army being raised to assault and capture Louisbourg, Nova Scotia with the grade of Ensign in 1745. After he completed his service during King George's War 1722 to 1748, he became a merchant in New Haven, Connecticut. He married his wife Mary Saltonstall in 1750. They had eight children.

At the start of the French and Indian War (1754 to 1763), Whiting was appointed Lieutenant Colonel of the 2nd Connecticut Militia Regiment. During the (Bloody Morning Scout) he was in command of the rear section of Colonel Ephraim Williams column and took command of the column after Colonel Williams was killed; organizing and directing an excellent retrograde movement that saved the majority of the column.

He was promoted to full Colonel, Connecticut Militia in 1756. In 1757 his regiment was at Fort Number Four on the Connecticut River. After the war he served in the Connecticut General Assembly until his death on 9 April 1771. He is buried in the Grove Street Cemetery, New Haven, Connecticut.

Colonel John Goodridge or Gutridge (both spellings are used) is truly a shadow in the mist. He was known to George Washington as his name appears twice in Washington's papers. It is possible that General Washington was referring to a younger family member with the same name. Records are from family genealogies where members of that family married into his family. He was apparently from Massachusetts – the Amesbury or Newbury area. Records show he was a Colonel of a Massachusetts Regiment and died in service. He did not die at Lake George and the date, details and location of his death are unknown as is his date of birth and life in general.

Lieutenant Colonel William Eyre, 44th Regiment of Foot (The Irish Regiment, so named as it was raised solely and totally in Ireland) arrived in North America in 1755. He was an Engineer and assigned to the construction of Fort Edward or Fort Lyman as it was originally named. He designed Fort William Henry and acted as Artillery Officer during the Battle of Lake George – September 8, 1755. After the battle he stayed at Lake George and supervised the construction of Fort William Henry and served as Commander of the fort until 29 March 1757, when he took detachments of the 44th Foot and 48th Foot to Albany to be moved to and involved in the 1758 siege of Louisbourg in Canada, He was promoted from Captain to Major in 1757 and Lieutenant Colonel to the 44th on 31 October 1759. He left for home (Ireland) in 1760 but never got home as the ship he was on sunk with all hands due to a heavy storm in the Irish Sea.

Citizen soldiers. They defeated French Professionals.

REFERENCES:

Fort George Advice, Fall 2003, King Hendrick

Fort George Advice, Fall 2004, Ephraim Williams

Fort George Advice, Fall 2007, Major General Seth Pomeroy

Fort George Advice, Spring 2008, A Second Look at the Battle of Lake George

Blodget'd Prospectus – Plan of The Battle Near Lake George, 8 September 1755

New York State Military Museum on Line Search.aol.com/aol/search? Colonel John Goodridge

Archieve.org/strm;Rhode Island in col. Rhode Island in the Colonial War

Search.aol.com/aol/search's Colonel Moses Titcomb

Wiikipedia, the Free Encyclopedia – Phineas Lyman

Ctheritafe.com/encelopedia'topical surveys/ctatwar,htm-Connectocut at War

The Montresor Journals, New York Historical Society – 1881

The Summer Paradise in History, Warwick Stevens Carpenter, The Delaware and Hudson Co., Albany, New York – 1914

New France and New England, John Fiske, Heritage Books, 1997

Relief is Greatly Wanted, The Battle of Fort William Henry, E. J. Dodge, Heritage Books, 1998

Muster Rolls of New York Provinical Troops, 1755 – 1764, New York Historical Society – 1892

Sir William Johnson's Papers, University of The State of New York, Pg 860, 1921

Wikipedia, The Free Encyclopedia – Nathan Whiting

Freeman, Freeholders and Citizen Soldiers, an Organizational History of Colonel Jonathan Bagley's Regiment 1755 to 1760, Brenton C. Kemmer, Heritage Books, 2004

History – St. John's Lodge Number One, Portsmouth, New Hampshire – On Line 2008

JARED SPARKS

Few people unless they are within the venue of early American literature, early American history or members of the Unitarian Church will recognize the name Jared Sparks.

He was one of nine children born to Joseph and Elinor (Orcut) Sparks of Willington, Connecticut. He was born on May 10, 1789. When he was six years old he went to live with relatives in Camden, New York to ease the financial burden at home, returning home in 1805.

He was quite intelligent and was known as "the genius" in grammar school. His interests were literature and history and he later developed an interest in astronomy. He worked at age 18 as a journeyman carpenter and school teacher. He started to study mathematics and Latin at age 20. He was able to obtain a scholarship to Phillips Exeter Academy. He was admitted to Harvard in 1811 and left in 1812 due to financial reasons and tutored a family in Maryland, later returning to Harvard and graduating in 1815. While attending Harvard Divinity School from 1817 to 1819, he tutored geometry, astronomy and natural history.

Leaving Harvard, he became a minister at the First Independent Church (Unitartian) in Baltimore, Maryland. As his feelings for the ministry were less then strong, he resigned his position in April 1823 returning to Cambridge, Massachusetts.

It was at this point that his life lead him into the literary world as an owner, publisher and writer. He is probably best known for the twelve volumes of "The Writings of George Washington", which he completed between 1834 and 1837. His literary accomplishments are numerous covering early American history. His discourse with Robert Peel of Great Brittan over the release of some historical papers that Great Brittan had concerning the Revolutionary War has become a rather lengthy discourse by other individuals.

How does Mr. Sparks become of interest to the readers of this newsletter?

In 1830 Jared Sparks traveled through the Lake Champlain, Lake George area making notes and some sketches. His notes on Fort George are as

Jared Sparks by Rembrandt Peale, Harvard Art Museum, Fogg Art Museum, Harvard University Portrait Collection, bequest of Lizzie Sparks Pickering, H244, Cambridge, Massachusetts.

Follows: "Fort George is a third of a mile from William Henry, and on much higher ground. It has high wells of stone, surmounted by earth. It is a very irregular structure, about 70 yards long by 50 yards wide, in its extreme dimensions. The ditch must always have been shallow, and the masonry of the walls was clumsy and imperfect. It seems to have been intended chiefly as a depot for provisions and other supplies, for the interior is much taken up with subterranean apartments strongly walled and covered over. It could have held but few men.

Water was procured by a covered way from the river, 2 or 300 yards distant. A hill within ¾ of a mile commands this fort. It is surmounted by a fortification which the people now call Fort Gage. The foundation still seems of a very large warehouse, doubly used in the Revolution as a temporary storehouse for supplies for Burgoyne's army, which were brought up Lake George. The walls of Fort George as they now stand are from 15 to 20 feet high with a covered way to the lake." His diagram resembles a crystal tear drop pendent from a crystal lamp with the gate at the lower end of the pendent opening toward the lake.

After this trip Jared Sparks went on to be involved with his historical writings, working for and on behalf of the Unitarian Church, education in general and President of Harvard University from February 1, 1849 to February 10, 1853. He married Francis Anne Allen in 1832. They had a daughter. Francis passed away in 1833. He remarried on May 21, 1839. His second wife was Mary C. Silsbee. They had four children.

Jared Sparks passed away on March 14, 1866.

REFERENCES:

Pell Research Center – Fort Ticonderoga, Ticonderoga, New York

The Illustrated Columbia Encyclopedia, Vol. 19, Page 5848

University of Georgia Library System Internet

ABC-CL10 on Line

Harvard's Unitarian Presidents, edited by Herbert F. Vetter @
http://www.harvardsquarelibraty.org/HVDpresidentssparks.php

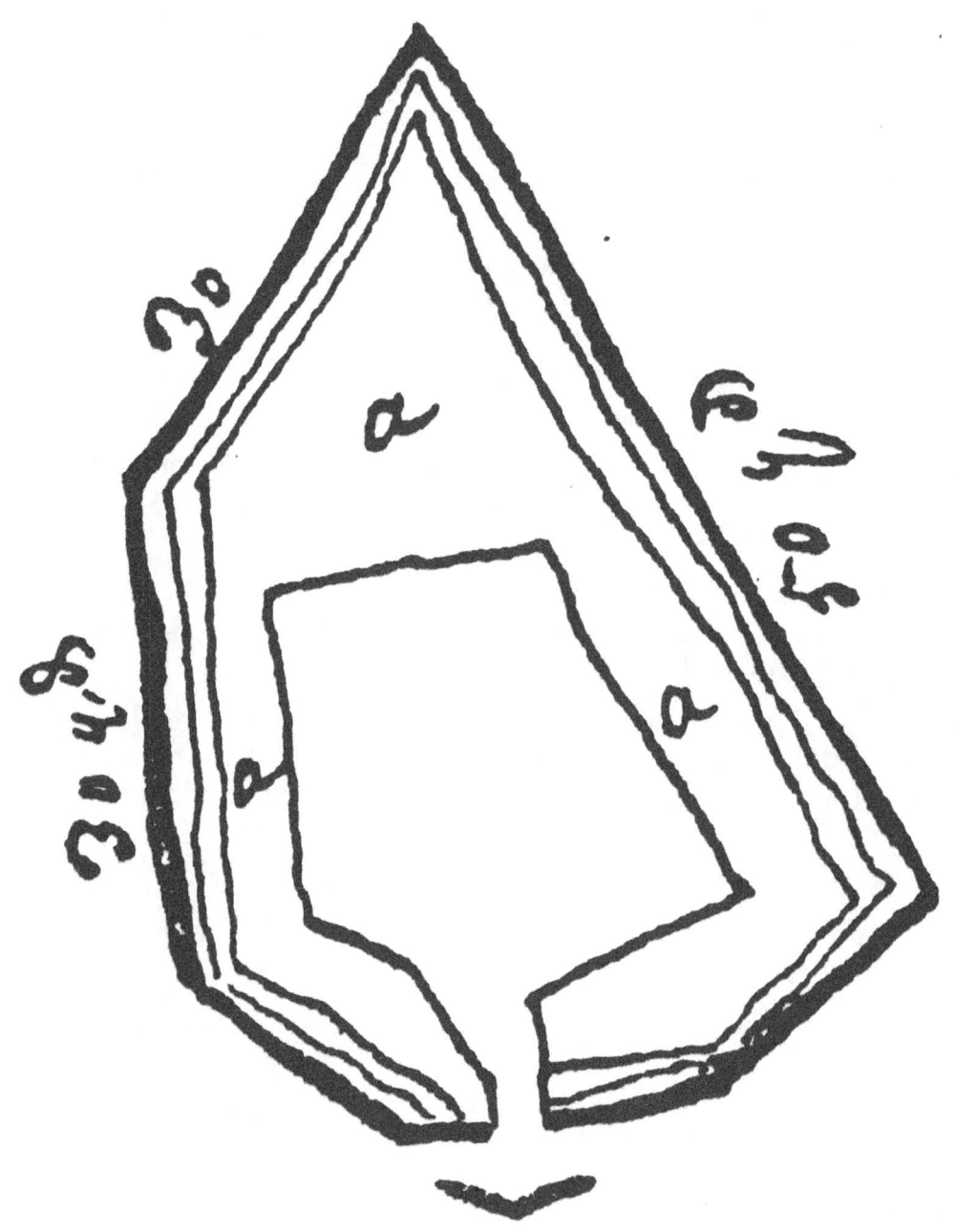

Jared Sparks' sketch of Fort George, August 1830
MS Sparks 128, Houghton Library, Harvard University,
Cambridge, Massachusetts.

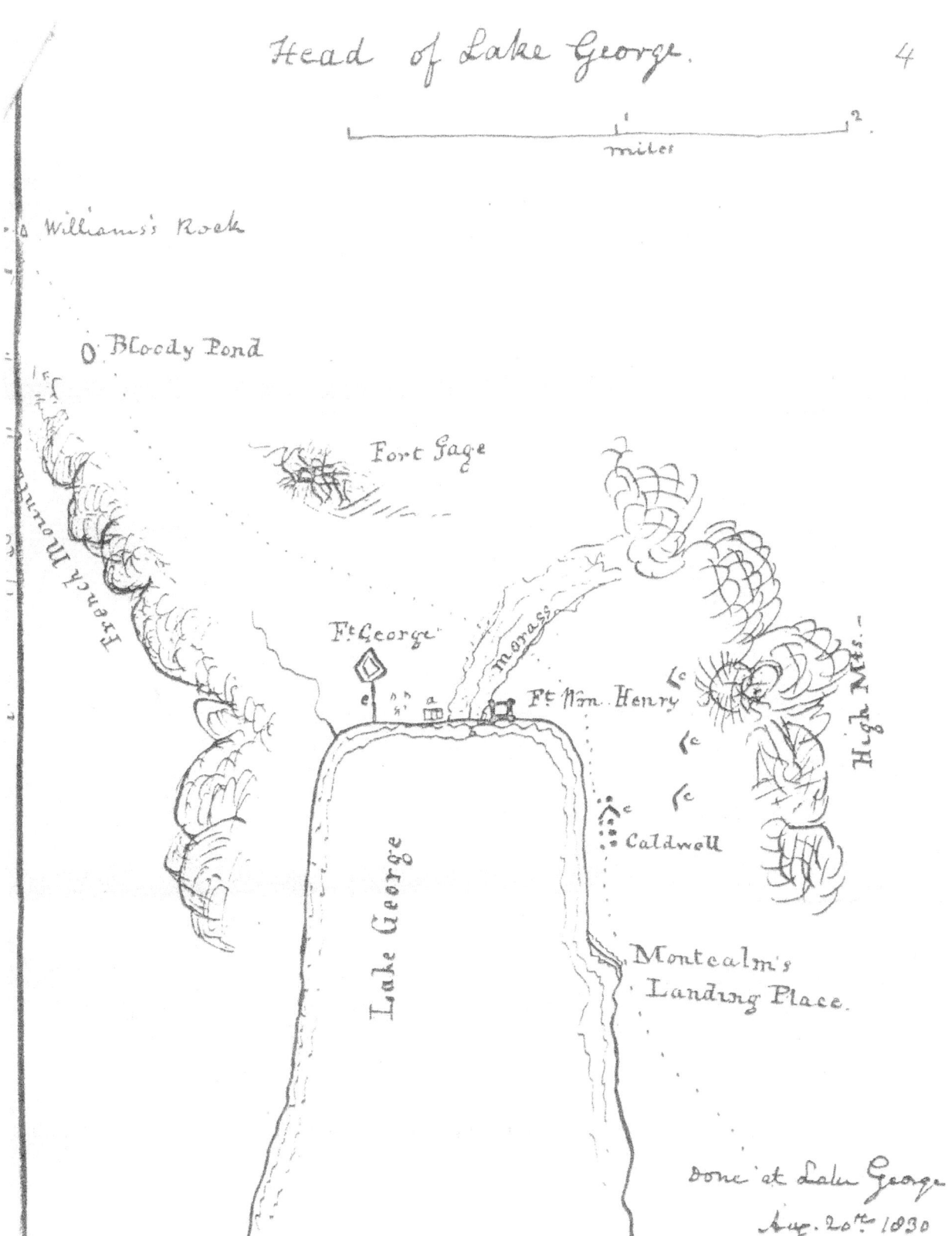

Head of Lake George.

The 253 Year Old Mystery

History is a mystery until five questions have been asked and hopefully answered - Who, What, Where, When and Why.

The referenced mystery is known and unknown, answered and unanswered and may never be completely solved as some questions will remain unanswered.

French Mountain is sited so that both it's western and eastern slopes end on it's north in Lake George with it's eastern slope footing on the valley that's runs north and south from Assembly Point/Cleverdale on Lake George to Glens Falls. The mountain's western slope foots on the narrower valley that runs north and south from Lake George Village to Glens Falls.

On 22 May 1756, Captain Gaspard De Lery at Fort Carillon entered in his diary that Captain Colombiere had left the fort with a force of 403 La Marine, Militia and Indians to raid Fort Edward. Colombiere returned on the 29th with four prisoners and three scalps. Colombiere had proceeded south by way of South Bay, Wood Creek and the east slope of French Mountain.

Somewhere on the mountain the French encountered a Provencial patrol commanded by a Lieutenant Thomas Brooks of Bagley's Regiment from Massachusetts that was stationed at Fort Edward. As to the specifics of what actually happened, it is unknown. Shots were fired and a Moses Burlong an interpreter with the French was killed as well as Lieutenant Brooks. The patrol was probably a section in strength or six men plus the Lieutenant. According to De Lery's diary at least two other Provencials were killed.

In 1999 I received a request from a descendent of Lieutenant Brooks to see what I could find out about the incident. The request was occasioned by the descendent's daughter finding a gravestone in the "Wilsey Cemetary" in the village of Thurman, Warren County, New York. The stone was engraved "Thomas Brooks, Slain in The French War near The French Mountain". The daughter felt that this was a memorial only as no dates were reflected.

I searched the militia records of New York and found that the records for

1756 were missing and had been since 1862. I contacted a friend who was and is involved with the reenactment group that portrays Bagley's Regiment and he checked and found a Thomas Brooks as an Ensign in Captain Samuel Brooks Company in Bagley's Regiment in September 1746. There was no further information on Thomas Brooks until the appearance of a newspaper article in the New York Mercury dated Monday. 23 August 1756.

The article related that during the questioning of three French prisoners on or about 25/26 July 1756, it was learned that Lieutenant Thomas Brooks had been killed some five or six weeks earlier on approximately 26/27 June. Part of the mystery; DeLery's diary puts Colombiere outbound from Fort Carillon on it appears the 21st of May and back on the 29th. The dates could well be in the month of June and the entries not properly dated or misread.

Part two of the mystery. Captain Colombiere. It has taken ten years to identify this man. He was a native born Canadian - Louis-Francois de La Corne.

He was born at Fort Frontenac, Canada in 1703. He joined La Marine asa Cadet in 1719, promoted to 2d Ensign in 1722, Ensign in 1727, Lieutenant in 1738 and Captain in 1744. He married in 1744. He was in the fur trade, saw action in Arcadia which earned him the Cross of Saint Louis. That was the Medal of Honor or Victoria Cross of that time. He was in the Battle of Quebec and after the defeat was ordered to France by the British and sailed with his brother St. Luc de La corne on 15 October 1761 on the Auguste which sank in a storm off St. Breton Island in November 1761. He did not survive the sinking. His brother St. Luc survived and is remembered in history as being in charge of General Burgoyne's Hyenas.

Are the questions answered?

WHO: Lieutenant Thomas Brooks vs. Louis-Francois de La Corne (Colombiere);

WHAT: Small unit (patrol) action;

WHERE: East slope of French Mountain.

WHEN: Probably June 1756 (Possibly May 1756).

WHY: It was another episode in a war.

UNANSWERED QUESTIONS? You can make up your own. One question that will always be there - where is Lieutenant Brooks body? For all these years he has been part of French Mountain. As good a spot as any.

His descendent family have been provided the information concerning the Lieutenant.

REFERENCES:

Diary of Captain Gaspard DeLery, New York Historical Society
New York Mercury of Monday, 23 August 1756
Harvey Pullen, Lakewood, Washington - Descendent
Carol Reynolds - Descendent
Vol. I, The History of Robert Rogers Rangers, Burt G. Loescher, Pg 320, Heritage Books, 2001 Reprint of 1945 edition
Historical Biographies, Nova Scotia, Louis La Corne (1703-1761)
LaMarine, Gallup & Shaffer & Lee, Heritage Books, 1992
Luc de La Corne - Wikipedia, the free encyclopedia
Dictionary of Canadian Biography online, Vol. III (1741-1770), Vol. IV (1771-!800)
La Corne de St. Luc du Chapts, users us internet.com/dfnels/lacorne-zip,htm
The Wreck of the Auguste, National Historic Sites Parks Service, Environment Canada
Brenton C. Kemmer, Educator, Historian, Author, Houghton Lake, Michigan

THE FRENCH AND THE LAKE

It would be unrealistic to take the position that the French had only limited knowledge of Lake George. Cabot had identified New France in 1497 and the French had been in and out of New France since 1504. They established a permanent settlement at Quebec in 1608 and stared serious exploration to the west and north using the St. Lawrence River as a guide or base line.

By 1609 they had found and named the Richelieu River and Samuel Champlain had gained access to the lake he named after himself. He may have learned of the existence of Lake George from his Indian guides but never went further then to name the portage "La Chute". He and his guides did get into a battle with some Iroquois which Champlain won because of his use of fire arms which resulted in the death of three Iroquois leaders and the pure hate of the Iroquois toward the French.

The Iroquois wanted part of, if not all the fur trade that existed between the French and northern and western tribes that had existed since 1607, possibly as early as 1605. Raids by the Iroquois had gotten so many and deadly that by 1683 the Companies Franches de La Marine were sent to New France.

In 1687 a force of over 3000 militia and La Marine headed down the Richelieu into what was known as the "Wilderness" in an attempt to punish the Iroquois. They found few Indians, but extensively destroyed crops the Indians were raising. In 1689 the Iroquois returned the raid with the massacre at Lachine near Montreal. They burned 56 houses, took 90 prisoners, killed 24 and roasted five children while Montreal watched. They then breached the gate at Montreal, destroyed large amounts of goods and killed several individuals.

France retaliated by encouraging their Indian allies to attack Iroquois villages and British settlements as they felt the British were behind the attack on Lachine. They raided Schenectady, New York using the Richelieu, Lake Champlain and Lake George as the most direct route. A year later the Iroquois attacked Montreal which had been settled in 1642 and Trois-Rivieres which had been settled in 1634.

The French were able to trade with the Iroquois intermittently and had established a Blockhouse trading post on the Hudson near present day Albany, New York in 1640. They were of the opinion that the Hudson or "Grand River" as it was called, conjoined with the St. Lawrence at some point. They were looking for the "Northwest Passage".

In attempting to expand New France and to offset the growing British/American settlements and individual incursions, the French built a fort on Lake Champlain as well as at least two more on the Richelieu. Fort St. Frederic was also the site of a large French village. It was a base of operations from which they and their Indian allies could raid into New England and northern New York using Lake Champlain, Lake George and Otter Creek as lines of travel. The fort was built in 1733, although a small fortified post had existed at Chimney Point located just across the lake from the fort location from 1731 to 1733.
The British would refer to the fort as Crown Point.

The French started to take critical notice of the British incursions with Arcadia, the attack on Louisburg - both in 1745, Braddocks attempt to attack Fort Duquesne in 1755 and Governor Shirley's plan to attack Fort St. Frederic in 1755. The French sent Baron Dieskau to Fort St. Frederic. From there he was to attack Fort Lyman (Fort Edward) on the Hudson. He proceeded south by Lake Champlain, South Bay and Wood Creek. While in route he learned of General Johnson's camp on Lake George. When his Indian allies learned of the cannon at Fort Lyman, they refused to attack, but did agree to attack the camp at Lake George. Baron Dieskau was defeated in that battle. (See Fort George Advice - Spring 2005). The immediate response of the French was to start construction of Fort Carillon in October 1755. The British referred to it as Fort Ticonderoga.

In March 1757, Pierre Francois Vaudreuil attacked Fort William Henry which had replaced General Johnson's camp. The attack was repelled. In August 1757 General Montcalm attacked Fort William Henry, forced it's surrender and the total destruction of the fort.

Samuel de Champlin was the first of many Frenchman to be directly or inderctly involved with Lake George. He was born it is believed about 1580 and is known as "The Father of New France". He was a navigator, cartographer, soldier, explorer, administrator and chronicler. He died on December 25, 1635. He was buried in Quebec, but his grave site was destroyed by fire in 1640 and its location is no longer known.

Father Isaac Jogues traveled Lake George twice. The first time in 1642 when he was captured and tortured by the Mohawks. He was able to escape with the help of Dutch merchants. He returned to France and then back to New France. He traveled Lake George again in 1646. It was during this trip that he named the lake "Lac du St; Sacrament". He was killed by the Mohawks on 18 October 1646. It is believed that his grave is located near Auriesville, New York. He was young, having been born 10 January 1607.

Pierre Francois de Riguad, Marquis de Vaudreuil-Cavagnal was born in New France on 22 November 1698. He was the son of Phillippe de Riguad Vaudreuil, the Governor General of New France. He was appointed governor od Trois-Rivieres in 1733 and in 1742 became governor of French Louisiana, serving from 10 May 1743 to 9 February 1753. He became governor of New France in 1755. He ordered the attack on Fort William Henry. He died in Paris, France on 4 August 1778.

Jean-Armand Dieskau, Baron-de Dieskau we have written about before (see Fort George Advice - Spring 2005). He was born in 1701 in Saxony and died in 1767 at Suresnes, France. He was in charge of the French and Indian force that attacked General Johnson's camp at Lake George in 1755. He was wounded, captured and his force defeated.

Louis-Joseph de Montcalm was born 28 February 1712 near Nimes, France, of a noble family. He entered the army in 1722 as an Ensign in the Regiment d'Hainault. On the death of his father he became th4 Marquis de Saint-Veran. He became a Captain in 1729. Served in several European conflicts and was promoted to Colonel in 1743 and awarded the Order of Saint Louis in 1744. Promoted to Brigadier in 1746. In 1749 he was awarded the opportunity to raise a new regiment; the Regiment de Montcalm, was a cavalry regiment. He was promoted to Major General in 1756 and sent to New France. He captured and destroyed Fort Oswego in 1756 and did the same to Fort William Henry in 1757. He defeated General Abercrombie at Fort Carillon (Ticonderoga) in 1758. At the Battle of The Plains of Abraham at Quebec, New France on 13 September 1759 he was mortally wounded, dying 14 September 1759. He is buried in the cemetery of the Quebec General Hospital.

Louis Antoine de Bouganville was born 12 November 1729. He studied law, but left the profession and joined the army in 1753. In 1755 he was sent to London as a secretary to the French Embassy and was made a member of the Royal Society. He arrived in New France in 1756 as a Captain of Dragoons and Aide-de-Camp to General Montcalm. He was at Oswego in 1756 and Fort William Henry in 1757. He wrote in his journal how when he delivered a message from General Montcalm to Colonel Monro, that the Colonel "Returned many thanks for the courtesy of our nation and protested his joy at having to do with so generous enemy". He also wrote to his mother regarding the massacre at Fort William Henry, "Some who would call themselves French took part in the massacre". He was wounded at Fort Carillon (Ticonderoga) in 1758 and returned to France. He was awarded the Cross of Saint Louis and returned to New France in 1759 as a Colonel. He was at Quebec, but saw no action as his assignment put him upstream toward Montreal. He wrote in his journal "It is an abominable kind of war. The very air we breathe is contagious of insensibility and hardness.". He returned to France in 1761 and acted as a diplomat from 1761 to 1763, helping to negotiate the Treaty of Paris. After the war he began exploring the world, particularly the Pacific Ocean area. He wrote a two volume set of books about his travels. He died 31 August 1811 in Paris, France.

Francis de Gaston, Chevalier de Levis was born 20 August 1717 near Limoux, France. He joined the army at age 15. He fought in numerous European conflicts and met and became friends with Montcalm. He was sent to New France in 1756. He was at Fort William Henry in 1757 in charge of the force that came down the Indian trail on the west side of Lake George. He was in charge of the French positions to the west and south of the fort. He was at Fort Carillon (Ticonderoga) in 1758. After Montcalm was killed, he was appointed Commander of French Forces in North America. He returned to France in 1761 and the King made him the Duc de Levis. He was appointed governor of Artois. He was promoted to Marshal of France in 1783. He died in 1787 in Arras, France.

Francois-Charles De Bourlamaque was born in Paris in 1716. It is believed he was of Italian descent. He entered the army in 1739 in the Regiment du

Dauphin. He rose to 2nd Lieutenant in 1740, Adjutant in 1745 and Captain in late 1745. He is designated a military engineer. In 1755 he was awarded a monetary award for his two years of work improving the infantry drill book.

He came to New France in 1756 as a Colonel, third in command behind Montcalm and Levis. He was awarded the Cross of Saint Louis in 1756. He was involved with the attack on Fort Oswego and Montcalm charged him with the direction of the siege at Fort William Henry in 1757. He finished the year at Fort Carillon (Ticonderoga). He was there in 1758 and wounded during the attack. He recovered at Quebec. He was at Fort Carillon in 1759 and left attempting to blow it up. Only one bastion was blown up. He blew up Fort St. Frederic totally destroying it. He spent the rest of 1759 fighting delaying actions against the British.

He returned to France in 1761 and was promoted to Commander in the Order of Saint Louis. He was sent to Malta in 1761. In 1763 as a Major General he was appointed governor of the colony of Guadeloupe where he died in office in 1764.

Little is known about this officer, Lieutenant Colonel Francois Le Mericer. There is evidence that he was in Illinois at Cahokia, but no date. He was at Fort William Henry twice in 1757. Her was there during the attack in March 1757 where he acted as the go between Vaudreuil and Major Eyre. He was there again in August as Montcalm's artillery officer. He asked Montcalm to deliver his compliments to the commander of the Fort artillery for performing his job so well, which Montcalm did. Nothing further could be found on this officer.

Joseph Marin De La Malgue was born in 1719, the son of Lieutenant Colonel Paul Marin De La Malgue, colonial militia. Joseph joined La Marine in 1732 and was sent west where he spent the next thirteen years in what is now Michigan and Wisconsin. His influence with the tribes was so great that he allied 22 tribes with the French. He was promoted to Ensign in 1750. In 1751 he was in Quebec. 1752 saw him back in the west. Back to Quebec in 1754 and then back west until 1756.

In 1756 he was in the Lake George area with his 150 Minominee warriors.

In 1757 he was in the area of Fort Edward where he wiped out a 10 man patrol and 50 man guard. He rescued Captain Israel Putman, one of Rogers Rangers from the Huron in 1758. He was promoted to Captain in 1759 and captured the same year. He was sent to France, but returned to New France in 1762 and was captured again and returned to France. He was promoted to Lieutenant Colonel in 1773 and died in Madagascar in 1774. Montcalm referred to him as "Brave but stupid".

Claire-Louis-Francois de La Corne was born in 1703 at Fort Frontenac (present day Kingston, Ontario, Canada). His full name was Louis-Luc (Chevalier) de La Corne de Chapt Seigneur de Terrebonne. He became a Cadet in La Marine in 1719, 2nd Ensign in 1722, 1st Ensign 1727, Lieutenant in 1738 and Captain in 1744. From 1747 to 1749 he was in Arcadia. He was awarded the Cross of Saint Louis on 10 May 1749. He was recalled to Quebec in 1750. He was out west from 1753 to 1756. He returned east in 1756 and patrolled the St. Lawrence Valley from Fort Carillon (Ticonderoga) and the portage (La Chute) during May, June and July, scouting and raiding in the Lake George area. His activities from late 1756 to 1760 are not really known. He did have an active interest in the fur trade which his family was heavily into. He was in Quebec in 1760 and possibly 1759. He was ordered to France by the British after the surrender and left in 1761 on the Auguste which sank off St. Breton Island, New France, in a heavy storm at sea. He did not survive the sinking. He was also known as "Columbiere" and his military company was known as "Company Columbiere".

Luc de La Corne, Sieur de Chapt de St. Luc was born in 1711. He became an Ensign in La Marine in 1742, Lieutenant in 1748 and Captain in 1755. He was out west in 17423 and 1743. In 1745 he and his brother Louis-Francois were in the raiding party that destroyed Saratoga, New York (present day Schuylerville, New York). In 1746 he was at Fort St. Frederic (Crown Point) from January to April conducting raids in the New England area. In 1755 he was one of the officers attached to the Indians in Baron de Dieskau's force during the battle in September of that year.

In 1758 after General Abercrombie's failed attack on Fort Carillon (Ticonderoga), La Corne attacked a supply train on its way to Fort Edward. The train was totally destroyed with LaCorne taking 64 prisoners, 80 scalps and destroying approximately 250 head of oxen. He returned to Montreal in

1759 and Governor Vaudreiul made him a Chevalier of Saint Louis in that year. He was at the battle of Sante-Foy in 1760 and was wounded. He was like his brother Louis-Francois ordered to France by the British in 1761. He sailed on the Auguste as well, but survived the storm at sea and sinking of the ship. He got to St. Breton Island and walked from there to Quebec in a four month trip. The British allowed him to stay.

He is best known for being in charge of General Burgoyne's Indians known as "Burgoyne's Hyenas" by the British Parliament. Governor Guy Carlton of Canada considered LaCorne "A great villain and as cunning as the devil". He died on 7 October 1784 and is buried in the Chapel of Saint-Anne in the church of Notre Dame, Montreal, Canada.

New France has been Canada since 1763 and that country's knowledge of Lac Du St. Sacrament or Lake George goes back to at least 1609. Maybe, just maybe they have a better claim to the lake then we do – spiritually.

REFERENCES:

The Summer Paradise in History by Warwick Stevens Carpenter, The Delaware and Hudson Company, Albany, New York, 1914

La Marine, Andrew Gallup, Donald F, Shaffer, Joseph E. Lee, Heritage Books, Bowie, Maryland, 1992

Relief is Greatly Wanted, The Battle of Fort William Henry by Edward J. Dodge, Heritage Books, Bowie, Maryland, 1998

New France and New England by John Fiske, Heritage Books, Bowie, Maryland, 1997

THE Illustrated Columbia Encyclopedia, Columbia University Press, New York and London. Volume 10, Pg 2981

Luc de La Corne – Wikipedia, the Free Encyclopedia

Dictionary of Canadian Biography Online Volume III (1741-1770), Volume IV (1771-1800)

Historical Biographies, Nova Scotia: Louis La Corne (1730-1761)

The Wreck of The Auguste, National Historic Sites Parks Service, Environment Canada

Francis de Gaston, Chevalier de Levis, Wikipedia, the Free Encyclopedia

Louis Antoine de Bougainville, Wikipedia, the Free Encyclopedia

Joseph Marin de la Malgue, Wikipedia, the Free Encyclopedia

Francois-Charles De Bourlamaque, Dictionary of Canadian Biography Online

Pierre Francois de Rigaud, Marquis de Vaudreuil-Cavagnal, Wikipedia, the Free Encyclopedia

Louis-Joseph de Montcalm, Wikipedia, the Free Encyclopedia

Jean-Armand Dieskau, Baron de Dieskau, Dictionary of Canadian Biography Online

Issac Jogues, Wikipedia, the Free Encyclopedia

Samuel de Champlain, Wikipedia, the Free Encyclopedia

Francois Le Mericer, AOL Search, Notes on Old Cahokia, AHOKIA, Illinois

Beaver Wars, Wikipedia, the Free Ebcyclopedia

History of Canada, Wikipedia, the Free Encyclopedia

ABOUT THE AUTHOR

This is Ed's second book about the French and Indian War. In the first book "Relief is Greatly Wanted – The Battle of Fort William Henry", he wrote about the battle that occurred at the Fort in August 1757.

In this book he writes about the first major battle in North America which occurred some 24 months prior in September 1755 within a few hundred yards of the site of the future Fort William Henry.

Ed writes about the history of what he calls "The Door" – Lake George; the principles on both sides of the battle and what happened during that September 1755 battle. The book is referenced chapter by chapter with maps, photographs, paintings and bio's on some of the persons involved; showing them to be the same as we are. Same language, same problems, willing without question to fight for family, home, country and their individual beliefs.

Ed Dodge has written articles for the "Bulletin" published by the Military History Society in England as well as historical groups in American covering not only the French and Indian War, but WWI and WWII. He has also written monograms on Alzheimer's which has afflicted his wife.

He served in the Marine Corps and Air Force earning Good Conduct awards from both services as well as the National Service Defense medal with a 2nd award, the United Nations Service award and the American Legion Award as Outstanding Plebe in his Army ROTC unit. As a youngster he earned his Explorer Gold Award in the Boy Scouts (The same as Eagle Scout) and as an adult leader was awarded the Bronze Big Horn award for service to the BSA Exploring program.

Ed is retired from the insurance business and security business, making his home in Springfield, Illinois but referring to Massena, New York as home.

BIBLIOGRAPHY

The bibliography for each chapter in this book is listed at the end of the chapter. There is no index as the information again is listed at the end of the chapter in all instances under the title REFERENCES.

www.ingramcontent.com/pod-product-compliance
Lightning Source LLC
Chambersburg PA
CBHW080601090426
42735CB00016B/3312